A Sniper's Conflict

A Sniper's Conflict

Monty B

Pen & Sword
MILITARY

First published in Great Britain in 2014 by
Pen & Sword Military
an imprint of
Pen & Sword Books Ltd
47 Church Street
Barnsley
South Yorkshire
S70 2AS

Copyright © Monty B 2014

ISBN 978 1 78346 220 9

The right of Monty B to be identified as the Author of this Work has
been asserted by him in accordance with the Copyright, Designs and
Patents Act 1988.

A CIP catalogue record for this book is available from the British
Library.

Typeset in Ehrhardt by
Mac Style, Bridlington, East Yorkshire
Printed and bound in the UK by CPI Group (UK) Ltd, Croydon,
CR0 4YY

Pen & Sword Books Ltd incorporates the imprints of Pen & Sword
Archaeology, Atlas, Aviation, Battleground, Discovery, Family
History, History, Maritime, Military, Naval, Politics, Railways, Select,
Transport, True Crime, and Fiction, Frontline Books, Leo Cooper,
Praetorian Press, Seaforth Publishing and Wharncliffe.

For a complete list of Pen & Sword titles please contact
PEN & SWORD BOOKS LIMITED
47 Church Street, Barnsley, South Yorkshire, S70 2AS, England
E-mail: enquiries@pen-and-sword.co.uk
Website: www.pen-and-sword.co.uk

Contents

Dedication

This book is dedicated to the everlasting memory of several friends who are no longer with us, who I still miss and who are in my thoughts daily. They have helped me to appreciate life in a much fuller way. After all, life is a gift and it is up to us how we use it.

Also to the men and women of our armed forces, all of whom have taken the Oath of Allegiance and choose to serve this nation with pride, professionalism and bravery, day in and day out until the job is done.

To our families for their everlasting love and support in all that we do.

To our American and Canadian brothers and sisters, who have served and are still serving. I have had the honour to serve alongside them on operations in Kosovo, Iraq and Afghanistan. Their courage and bravery are unequivocal and always have been, since the trenches of the First World War through to the battles of the Second World War, serving in Europe and the Middle East and in many other areas of conflict. This fellowship continues right up to the present day as we serve side by side in foreign lands.

To our NATO allies who have served with us with true military professionalism and dedication: soldiers' men and women from their respective countries, demonstrating so many deeds of bravery, compassion and sacrifice at all levels. Only they know of these deeds and actions, which matter to those of us who were there and we all remember them. Such sacrifice is just part of the price of serving in a country's armed forces, protecting and maintaining the freedoms that we value today.

Background and Foundation

The Core Functions of Reconnaissance: Find, Fix, Strike, Exploit

Sniping has not changed a great deal since the first man adopted shooting skilfully using an unconventional firing position compared to his fellow combatants who usually would fire from the standing or kneeling position in file among other soldiers. The early sharpshooter incorporated the use of natural elements such as foliage to help conceal him when firing from a chosen position anywhere on the battlefield.

An example of this is the back or backs position [supine with one shoulder raised and legs crossed to provide variable support to the rifle] that was adopted by British army sharpshooters in the era of the First World War, giving them a low silhouette and a comfortable position from which to be able to observe and fire. This could be sustained for some period of time with minimum movement. Another example from an even earlier period in time was the sitting position favoured by the early American sharpshooters during the War of Independence (1775–83) fighting in the dense forests of North America. Having thick foliage to observe and fire through acted as an excellent screen for them, providing some depth, shadow and backdrop (the sniper's allies) and of course these men were helped by not wearing the brightly-coloured uniform of the time, as sported by their military contemporaries.

Using a basic rifle and sight system already in service, the sharpshooter also mounted a very crude optical sight onto his rifle. This enhanced his observation of a specific singled-out target and the surrounding area if need be, enabling his first shot to be a well-aimed, well-placed, often fatal one.

Sniping is one skill that still requires a special kind of soldier with certain qualities. When I first began my training back in 1994 the basic principles were drummed into us, as were the history of sniping and all aspects of field training and marksmanship, instilling in me and others from that era a new passion for this military art and all that it encompasses. Qualities such as enduring patience, the hunter's instinct and acute powers of observation are just a few of the sniper's core skills, as quoted by Major Hesketh-Pritchard DSO, MC, the father of British army sniping from the First World War period.

In my opinion it is still the man with an above-average set of field craft skills, shooting ability and patience that excels in this area or, more importantly, survives. He does this by using his craft to the highest standard, enabling him to adapt to the ever-changing situation on today's modern battlefield: an open-minded, highly-trained and physically robust individual, competent in his personal skills as a soldier and those of a sniper.

The weapons, ammunition and optics we use today are far superior to those used by our forebears in past conflicts. So many improvements have been made over the years, both in war and peacetime, to the kit and equipment used by the sniper today. There are many available variants of the rifle itself, plus optical sights and spotting scopes, both for the military and civilian markets. They can even be built to individual specifications and requirements, custom-made for a specific task. There is the choice of the classic single-shot bolt-action rifle much preferred by hunters, for example, or a semi-automatic scoped rifle for use in high-tempo close operations by the military. Then the optical scopes themselves, now with a greater increase of magnification and clarity or sharpening of the target picture, giving pure clarity to the sniper's field of view of a target and area even at further ranges, depending on the shooter's skill.

The majority of modern rifle scopes have the capability to be used both day and night and in low-light conditions. The KN 203 night scope, for example, can be attached to the rifle by mounting it on the top of the day scope without needing to remove the day version from the rifle body itself. The KN 203 also has a daylight filter so it can be kept on the sniper rifle during the day if needed, depending on the situation or task you are in at the time. Other pieces of equipment used for observation by the sniper such as binoculars are crucial: a good lightweight, waterproofed pair with a military graticule pattern, coated lenses and a decent magnification option is ideal.

The choices available are abundant on the world market today and the same goes with regard to spotting scopes that can be used by the sniper pair or by other military specialists; also by civilians such as game-hunters, park rangers and anyone who requires enhanced vision of their subject of interest. A well-proven spotting scope is the Leupold model used by the British army which has a magnification of up to x40. Any such device that aids the sniper in his task is constantly being improved and developed.

Ammunition is just as important as improvements and advancements at all levels, and this may be manufactured for a specific task. For example, distance to the target, environment the round must travel through, type of target and impact or effect needed may all affect the end result, therefore pure attention to detail is required in choosing a round. This applies even more in the world of Special Forces scenarios and operations. The ammunition used by the green military is generally mass-produced for ease of production and delivery to the unit.

A piece of equipment worth mentioning is the pocket laser range-finder (PLRF-15C), used for measuring the distance to the target. It aids the sniper with pinpoint accuracy down to the nearest millimetre, eliminating human error. If the sniper had to make his own calculations, with just a little human error the drop or increase in range can make all the difference to a shot being on target, or a near miss alerting the target that someone is trying to engage him.

It is the man with the rifle who still has to get to his final firing position unseen, undetected and then conceal himself and his No 2, the spotter, in order to build up his firing position ensuring a clear line of sight and shot onto the target. All the time he is thinking on his feet and covering his ground sign [see Chapter 6] where possible on task, always thinking whether he has cover from view, cover from fire, a secondary firing position or positions if needed, and ensuring that his position has some sort of screen, depth, trapped shadow and backdrop. That is the position dealt with. He must then correctly estimate the range to the target within millimetres, now assisted by the modern laser range-finder.

Also he has to judge the wind accurately: this is a major factor that must be calculated correctly as it may have a big effect on the fall of shot. The speed of the wind must be considered: is it gentle, moderate strength, or moving up the scale to very strong? The direction of the wind is just as important as its strength. The round may be pushed slightly to the left or right and either up or down of the target, resulting in a miss. So again it is very important that we now have the use of modern wind meters, and the Kestral 3000 pocket weather meter is another valuable addition to the sniper's equipment.

Finally there is the decision to engage the target or targets: waiting, watching, seeing how the situation develops. To fire or not to fire, thinking of the bigger picture or plan? Has your mission changed or not in the time that you have been out on task? Also the rules of engagement are always at the back of your mind: just one wrong engagement may have a negative impact on an operation ('Hearts and Minds'); or a life may be in danger, in which case the threat must be taken out immediately.

Another sniper task and skill used both in conflict and on other operations is to observe and report: this is just as important as physically engaging a target and neutralizing the threat as it provides commanders with timely and accurate information. Sometimes a pair of human eyes able to observe a target or area is the only means of getting the required information at that point in time: this could be due to bad weather affecting the use of other types of surveillance devices. More importantly, a pair of snipers or a sniper screen may be able to infiltrate and get close to a target, without being detected or alerting the enemy, to obtain detailed information so that operations can be mounted against the threat using the most up-to-the-minute information available.

The skilled sniper must also be a communicator, having a good understanding of the signals/communications world; a trained medic to self-help his small team or pair who may get into trouble; and a competent navigator over any terrain in all weathers using and understanding mapping, air photography and the use of other modern navigational aids. Being trained in the use of other small-arms weapon systems and munitions, both in service and used by foreign armies, combined with a basic knowledge is a must. Finally he must be a man who always wants to learn and improve himself, both as a soldier and as a sniper, to be an asset to his comrades in this ever-changing and increasingly violent world.

So how did it all begin for me? I was influenced by my father, both my grandfathers and my great-grandfather, all of whom served in the British army. My father was Australian by birth and, going back to the First World War, on his side of the family a relative served in the 10th Company

Australian Machine Gun Corps and died of his wounds on 24 December 1916 at the Somme. As for me, all I knew from a very early age was that I wanted to be an infantry soldier.

I joined the British army in January 1987; yes, a long time ago. I did my training at the Junior Infantry Battalion in Folkestone, Kent. When I passed out of training I joined the 1st Battalion The Royal Hampshire Regiment in Tidworth who had just returned from a tour in Northern Ireland. This is where I met some truly inspirational characters and highly professional soldiers, who through the amalgamation in 1992 of my regiment with the Queen's Regiment, another fine infantry regiment of the line, then became the 1st and 2nd battalions, The Princess of Wales's Royal Regiment. There were some very experienced soldiers ready to teach and pass on their knowledge to us young pups and this was my starting-point.

My operational combat experience against a real enemy who actively targeted and engaged us using all means available to him didn't start until my first tour of Iraq when I was at the ripe old age of 33, and I was aged 38 for my first tour in Afghanistan. My previous operational experiences in Northern Ireland and Kosovo were not as intense or kinetic as what I was about to undergo serving in Iraq and Afghanistan. Running around with an L96 or an L115 A3 at that time in my military career, I was getting on in age, or at least I thought so. Still, age is just a number, as I am so often reminded by some of the troops with whom I served.

In the armed forces a great sense of loyalty, camaraderie and pride motivates people and that is what drives us to sometimes endure what we must to get the job done, often in extreme volatile situations in which anything could occur at any given moment as a situation develops or circumstances change. I think that all my training and the other tours prior to this plus experience gained over the years had prepared me in a way. A sort of stepping-stone, so to speak, leading towards what my colleagues and I were about to endure both physically and mentally in a foreign land, unfamiliar environment and a culture very alien to us.

Glossary of Terms

1 **PWRR**: 1st Battalion The Princess of Wales's Royal Regiment. An armoured infantry battalion stationed in Germany at the time with the Warrior armoured fighting vehicle.

2 **PWRR**: 2nd Battalion The Princess of Wales's Royal Regiment. Our sister battalion who were the theatre reserve battalion stationed in Cyprus at the time.

2IC: Second-in-command.

1st QDG: 1st Queen's Dragoon Guards. The regiment is part of the Royal Armoured Corps serving in the role of formation reconnaissance, also stationed in Germany at the time.

42 Commando: Royal Marines known as the Green Berets, 42 Commando is a commando light infantry battalion-sized unit and is part of 3 Commando Brigade. Like many other front-line units such as the Parachute Regiment and the Royal Anglian Regiment, they have been deployed to Afghanistan on several tours on Operation HERRICK along with various supporting units.

.338: Also known as the L115 A3 or Long Range Rifle (LRR). The British army's upgraded .338, single-shot, bolt-action sniper rifle, 8.59mm in calibre fitted with an adjustable telescopic scope with a magnification of x5-x25x56mm objective lens and, depending on the skill of the firer, can engage targets well over the distance stated in the training manual.

5.56mm: The standard rifle calibre ammunition used for the SA80 A2 individual rifle.

7.62mm: The standard belted calibre ammunition used for the general-purpose machine gun (GPMG).

105mm: The Royal Artillery's light field gun, 105mm in calibre, air-portable and can be vehicle-towed by in-service vehicles. Used mostly by the airborne forces and Royal Marine Commando units for close immediate support; able to fire high explosives and smoke and illumination rounds.

107mm: A high-explosive (HE) 107mm rocket used by the Taliban, left over from the Russian invasion. (The Soviets were in Afghanistan from 24 December 1979 until their last troops left on 15 February 1989.)

60mm mortar: Fires HE, illumination and smoke rounds and is the replacement for the 51mm mortar.

81mm mortar: The L16A2 81mm mortar can provide fire support with either HE, illumination or smoke rounds out to a range of 5,650 metres.

AK-47: Otherwise known as the Kalashnikov, a Soviet-designed assault rifle 7.62mm in calibre, used worldwide by insurgents and foreign armies.

Alternative firing position: A back-up position where a sniper can still observe and fire into the original target area. Where possible he should have several selected alternative positions that afford some degree of cover from view and protection from enemy fire.

ANA: Afghan National Army.

ANP: Afghan National Police.

AOR: Area of Responsibility.

AP: Armour-piercing; a jacketed bullet capable of penetrating light armour and body armour.

Apache: The AH-64, an attack helicopter manned by a crew of two and armed with a 30mm cannon, rockets and Hellfire missiles.

ASATS: Advanced Small Arms Targeting System, a ballistic data chart used by snipers. Once the range to the target and allowances for wind corrections and other environmental factors have been calculated, these corrections can then be applied to the elevation and deflection drums on the optical sight body, enabling the sniper to maintain the same point of aim on the target.

AT4: A single-projectile, shoulder-launched, disposable 84mm anti-tank weapon used by US and British forces in theatre.

Badged: The term used when a soldier has passed both phases of his basic sniper training to become a fully-qualified military sniper. This earns him the right to wear the highly prized sniper badge depicting two crossed rifles with an 'S' set above them.

Ballistics: The science of mechanics dealing with the flight, behaviour and effects of projectiles, particularly bullets and rockets, of which the sniper must have a sound understanding.

Bean bag: A sniper's aid for shooting, comprising a bag of either dry fine sand or dried peas waterproofed, put inside a camouflage material bag and sewn up or sealed with Velcro strips. Can be any size depending on the calibre of the rifle; used as a support and to adjust the angle of aim.

Bipod: A two-legged adjustable rifle support giving added stability and aiding accuracy during firing, enabling the sniper to maintain a firing position for sustained periods of time.

CASEVAC: Casualty Evacuation.

CASREP: Casualty Report.

Cheek piece: The .338 rifle is fitted with an adjustable cheek piece that can be customized to specific requirements for firing to the individual sniper, providing the correct eye relief and sight picture.

Chinook: A twin-rotor helicopter used by many NATO members for moving troops and equipment and for the CASEVAC of wounded soldiers.

Claymore: A remotely-initiated directional anti-personnel mine that may also be victim-operated. It is filled with ball-bearings and when sited correctly in a close environment can create a devastating effect.

CLP: Combat Logistics Patrol.

CLU: Command Launch Unit for the Javelin anti-tank missile. It can also be used for day and nighttime surveillance and is an excellent aid to the sniper pair.

Comms: Communications.

CQMS: Company Quarter Master Sergeant. An SNCO in charge of a team whose responsibility is to supply the company group with all necessary stores, kit and equipment.

CSM: Company Sergeant Major.

CVRT: Combat Vehicle Reconnaissance Tracked. Variants of light armoured tracked vehicles used in the role of reconnaissance, e.g. the Scimitar; mechanical support variants such as the Samson and Spartan vehicles; and medical support from the Samaritan field ambulance variant.

De-bomb: Soldier slang for the removal of ammunition from magazines, or any form of munitions from kit and equipment where it may be carried or stored.

Deflection drum: Once wind error to the target has been calculated and compensated for, the necessary wind correction can be applied directly to the rifle sight, enabling the sniper to use the normal point of aim onto the target.

DOP: Drop-off Point.

ECBA: Enhanced Combat Body Armour.

ECM: Electronic counter-measures. Man-pack or vehicle-mounted equipment designed to prevent the detonation of remote-controlled bombs and IEDs.

EHLS: Emergency Helicopter Landing Site.

Elevation drum: Raises or lowers the horizontal crosshairs of the sight to the required range needed to engage the target without having to change the point of aim.

EMOE: Explosive Method of Entry.

Endex: End of Exercise.

ERV: Emergency Rendezvous.

FAC: Forward Air Controller.

FOO: Forward Observation Officer. Trained to call in artillery to support the infantry or troops on the ground by providing HE and smoke or illumination rounds onto a specified target area.

FST: Fire Support Team.

GMG: Grenade Machine Gun 40mm. Used to provide immediate additional fire support with high-explosive grenades, the 40mm GMG can be vehicle-mounted and operated, making it a mobile weapons system.

GPMG: General Purpose Machine Gun. A 7.62mm belt-feed medium machine gun which in the light role can engage targets out to a range of 800 metres.

GSW: Gun Shot Wound.

H83: Ammunition container for the transportation and storage of ready-to-use ammunition or munitions components.

HE: High Explosive.

HESCO: 'Concertainer' multi-cellular defence barriers used for flood control and military fortification.

Hexi blocks: The Hexamine block is a smokeless solid fuel in tablet form used for heating water and food.

HLS: Helicopter Landing Site.

HMG: Heavy Machine Gun. A .30Cal or .50Cal Browning, either tripod- or vehicle-mounted.

ICOM: A lightweight portable radio receiver.

IED: Improvised Explosive Device. The name given to any unconventional explosive device made by the insurgents; it may be either remotely detonated or victim-operated.

L96: The L96 is an ageing, very reliable and proven sniper rifle with a single-shot bolt action, 7.62mm in calibre.

LASM: Light Anti-Structures Missile.

LOD: Line of Departure. A point on the ground from which, once left, a fighting unit is considered to be in a hostile environment.

Mark I Eyeball: unaided vision.

MC: Military Cross; a decoration awarded for acts of gallantry.

MERT: Medical Emergency Response Team.

MFC: Mortar Fire Controller. A mortar-trained and experienced NCO or SNCO who calls in mortars in a fire support role.

MSG: Manoeuvre Support Group. A mixture of specialist weapon systems such as the GPMG in the sustained fire role and the Javelin missile system.

NCO: Non Commissioned Officer.

OC: Officer Commanding.

OP: Observation Post.

ORBAT: Order of Battle.

Para Illum: Parachute Illumination. A hand-fired rocket flare attached to a small parachute, used to produce bright light for a limited time period.

Parallax: The displacement of an object relative to its background when viewed along two different lines of sight.

PB: Patrol Base.

PID: Positive Identification.

PLRF: Pocket Laser Range Finder. The quickest means available to the sniper pair to accurately determine the range and bearing to a target.

PMT: Police Mentoring Team. A team of experienced military police from all three services who train and advise the Afghan National Police.

PRR: Personal Role Radio.

QRF: Quick Reaction Force.

RAMC: Royal Army Medical Corps.

REME: Royal Electrical and Mechanical Engineers.

Reticle pattern: The pattern seen through the optical mounted rifle scope. It comprises a centre fine crosshair with a series of black dots (Mildots) that are evenly spaced along horizontal and vertical lines, used for measuring a target to help with estimation of range and wind drift.

RMP: Royal Military Police.

ROE: Rules of Engagement. A pocket-sized card issued to forces serving in a particular theatre of operations authorizing and/or placing limits on the use of force and the employment of certain capabilities.

RPG: Rocket Propelled Grenade, or RPG-7. A Soviet-designed shoulder-launched rocket system.

R&R: Rest & Recuperation.

RSM: Regimental Sergeant Major.

Sanger: A fortified purpose-built bunker or tower, part of a military location's security measures, manned by security forces and offering a field of observation and fire over a given area or sector.

Scimitar: Main vehicle used by the reconnaissance platoons of an armoured battle group armed with 30mm RARDEN cannon and 7.62mm coaxial machine gun.

SF: Sustained Fire (Role). Using the 7.62mm GPMG on a purpose-built tripod in conjunction with the gun optics increases range and allows for an increase in the rapid rate of sustained accurate fire.

Sight Picture: View of the target seen by a sniper when observing through optics.

SITREP: Situation report.

Snatch vehicle: A lightly-armoured heavy-duty Land Rover used on patrol.

SNCO: Senior Non Commissioned Officer.

Spot Code: A colour and number assigned to a specified point or location on the ground for ease of reporting; e.g. Yellow Three (a prominent road junction on the corner of the Al-Sadr building and the bridge).

Stand to: A traditional military term indicating raised awareness of a potential or actual threat. Also routine drill for soldiers at first and last light to mark the change in routine from day to nighttime.

Suppressor: A rifle attachment that muffles the sound signature and minimizes muzzle flash. Firing with the suppressor fitted must be practised and the data recorded because at various ranges the data will vary from that recorded without it.

SVD Dragunov sniper rifle: a Soviet-designed, reliable semi-automatic weapon. It is 7.62mm in calibre and magazine-fed, each magazine holding ten rounds of ammunition. Fitted with a PSO-1 x4 optical scope, it has an effective range of up to 800 metres.

UAV: Unmanned aerial vehicle. A remote-controlled aircraft used in hostile environments to gather intelligence with cameras or sensors and may employ weapons to neutralize a threat.

UGL: Under-slung Grenade Launcher.

VCP: Vehicle Check Point.

Viking: An armoured tracked vehicle, highly mobile over rough or wet terrain.

VP: Vulnerable Point.

Prologue

'Man Down'

Man Down at Yellow Three, QRF Stand To

On 18 April 2004 in Iraq, on Operation TELIC IV in the location of Al Amarah, we heard the sounds of rifle and automatic gunfire and the whizz of RPGs being fired in the early afternoon. The sounds of sporadic heavy dull explosions could be heard coming from the direction of Yellow Three and from the surrounding urban areas in and around what was known to us as the Al-Sadr building.

We had a call sign in that area on a routine mobile patrol in Snatch vehicles. The multiple commander and I stood to on hearing these sounds. Immediately the men sprang into action, grabbing their rifles, helmets and equipment and running to the vehicles from the rest area, which was a small paved area in the open with a few chairs out in the courtyard underneath two large palm trees in-between the portable cabins and main buildings. The drivers immediately jumped into their seats and started the Snatches up, revving them loudly and getting them ready to roll out at any moment. Ian and I then ran up to the operations room, en route getting our mapping out ready for a back-brief from the duty watch keeper.

Meanwhile over the radio in the background, information was being fed through from Dan to the operations room: he was the multiple commander of the call sign in contact. Everything was buzzing: it started to get busy with people coming in and out of the operations room, the guard commander was on his radio running around the corridors trying to get information from the Sanger sentries on what was happening outside the walls of our location and beyond. What could they see? What could they hear? Which direction was it coming from?

The call sign was caught up in a big contact and had received a casualty; heavy gunfire could be heard over the radio as well as the chaotic background noise of their surroundings coming over in the transmission. Dan sent his initial contact report calmly. Ian and I received our brief, short as it was with the information available to us at the time from ops and the Sanger sentries. We quickly came up with a plan. I remember at this point in time sweating like a maniac and thinking: 'This is it, this is real.' All the time rifle fire and small explosions could be heard in the background as we briefed the remainder of the call sign. Sweating from under my helmet onto the map on the front of the Snatch vehicle while briefing the boys, I was feeling slightly apprehensive and nervous at the same time.

By this point we could all hear what was going on outside the walls of CIMIC House over the noise of the powerful Snatch Land Rover engines. These were new sounds to us: pure chaotic violence, a ferocious fire-fight and the sudden thud of explosions. People could be heard screaming, shouting and crying, together with the sound of motor vehicles: just sheer volume of noise filling each man's ears. We could see from our location numerous plumes of dark billowing smoke rising, towering up into the clear blue afternoon sky in the direction that we were about to travel.

This was it. The Quick Reaction Force had been crashed out and we had a call sign needing our immediate help. One of our comrades was wounded out there somewhere in the heavily built-up urban area of Yellow Three. This was all that mattered: get to them, help them and bring them back, no matter what.

Chapter One

Iraq: Al Amarah

18 April 2004 CIMIC House, C/S Alpha 20 Bravo

The day started like most of the days that had passed before: nice and cool in the early morning, with a fresh light breeze coming in off the River Tigris into the compound. The sun was starting to break through the low-lying dark grey clouds from the night before: the dawn of another long hot sweaty day soon approached. The small garden in our compound and its plant beds were being watered and tended by two of the locally-employed civilians who were very elderly men, dressed in their long white dishdashas, who worked in our location maintaining the garden areas. The grass was kept immaculately short, very green and lush to look at, with a narrow concrete pathway cutting through the garden area itself en route towards the cookhouse.

People in the surrounding buildings around CIMIC House were going about their normal business as usual. A few of them were hanging out washing on the washing-lines on their balconies in the early morning, hoping to catch the sun before the wind and dust picked up as usually happened around midday. Others were putting out the rubbish from the previous evening onto the front doorstep or out of the side doors of buildings that gave onto the side streets and alleyways which were full of rubbish and waste and where some wild dogs were scavenging among the piles of refuse for food. A few children started to come out to play in the streets, skylarking about and chasing the dogs away from the rubbish and into the small alleyways that threaded in-between every building and solid wall. A small group of older boys had just come out of one of the alleyways and started to play football in the street in front of the main entrance to our location, kicking a half-inflated ball around and chasing it. The beginning of another day loomed ahead of us at the start of our tour and to us it was just like any other morning we had experienced thus far.

By this time the sound of several vehicles from outside the compound could be heard starting up their engines to take to the roads: soon all the roads outside would be busy with cars and trucks. Also a few of the local old men from the buildings opposite started to come out of their houses, greeting each other in the normal manner and talking outside in the street opposite our location just as on any other day that had passed before. The shop-owners started to open their shops: metal shutters going up to reveal the sheet-glassed shop fronts; commodities ranging from electrical items to fruit and vegetables set out on small tables on the street. The start of a normal day: just the usual routine, another day closer to the end of the tour had begun, or so we thought.

Today's plan of action was that a mobile patrol would be carried out by the sniper platoon multiple in the areas surrounding our location using two Snatch Land Rovers. Nothing mad: just a routine mobile reassurance patrol and as always a call sign would provide the Quick Reaction Force while the patrol was out on task. This was up to us, the reconnaissance platoon multiple. As usual, our drivers had to take over the Snatch armoured vehicles from the off-going call sign on that task. This entailed parading them and making sure everything to do with the running of each vehicle was functioning as it should be. Any faults had to be rectified there and then, or reported to the Company Motor Transport Rep. The multiple 2ICs also started to take over the QRF kit and equipment on the vehicles, again ensuring it was all there, serviceable and ready to use. This included spare ammunition, medical equipment, crowd control equipment, spare water jerry cans and the most important bit of kit, the ECM.

All this happened at a relaxed pace. Everyone else was checking their own personal equipment, packing the vehicles and going through breakfast. Ian being the multiple commander and myself the multiple 2IC, we went up to the operations room for a detailed brief on the patrol route, timings and marking-up of our maps and a back-brief on the previous night's events. We had all the information we required and briefed the remainder of the QRF. All the kit and equipment was good to go, the wagons were ready, and ourselves suited and booted with all weapons prepared. We had a runner in the operations room, and with radios on, we were fit to deploy if needed on task as the QRF.

The morning passed uneventfully, with the multiple having lunch in the small cookhouse overlooking the river. Shortly after this the sniper multiple deployed on their patrol task. It was early afternoon and the sun and temperature were at their highest; the wind started to pick up as it usually did at that time of day, resulting in an uncomfortably warm breeze.

Then all of a sudden we could hear strange sounds being carried along on the wind and for a few moments we had to stop and listen and try to recognize what these were. What we were hearing were the sounds of rifle and automatic gunfire and the very distinctive sound of intermittent high-explosive grenades mingling with larger more powerful detonations coming from rocket-propelled grenades as they joined in the evolving early-afternoon chaos. The sounds of sporadic heavy dull explosions could be heard from the direction of Yellow Three and the surrounding areas, where we had the sniper platoon multiple on a routine mobile patrol in Snatch vehicles. The multiple commander and I stood to the multiple on hearing these sounds. Immediately the men sprang into action, grabbing their rifles, helmets and equipment and running towards the vehicles from the rest area.

The drivers reached the vehicles first and jumped into their seats, starting the Snatches up: instantly the engines fired into life, revving loudly as they spluttered and spat out dark smoke from the exhausts to the rear of the wagons. Moments later they were purring away, ready to roll out at any minute. Ian and I then ran up to the operations room, en route getting our mapping out ready for a back-brief from the duty watch keeper. Meanwhile over the radio in the background, information was being fed through from Dan, the multiple commander of the call sign in contact, to the operations room. The place was a hive of activity, with people in and out of the operations room and the guard commander on his radio running around the corridors trying to get information from the Sanger sentries. What was happening outside the walls of our location and beyond? What could they see and hear?

Dan's call sign was involved in a big contact and had received a casualty. The chaotic background noise of their surroundings was coming over in the transmission, accompanied by heavy gunfire. However, Dan managed to send his initial contact report calmly. Ian and I received our brief, short as it was, with the information available at the time from the duty watch keeper and anything further from the Sanger sentries at the front main gate.

We quickly came up with a plan, although by this point in time I was sweating like fury and kept thinking: 'This is it, this is for real.' As we briefed the remainder of the call sign, I was sweating from under my helmet onto the map on the front of the Snatch vehicle. All this time rifle fire and small explosions could be heard in the background and while briefing the boys I couldn't help feeling slightly apprehensive and nervous all at the same time.

By now all we could hear outside the walls of CIMIC House were the sounds of a ferocious fire-fight and the sudden thud of explosions. It was just sheer loudness, with people screaming and the noise of motor vehicles. From our location numerous plumes of dark billowing smoke could be seen rising into the clear blue sky in the direction that we must take.

Immediately we got into the vehicles. Taff, my driver, started revving the engine: every time he put his foot down on the accelerator the V8 engine roared with power. I got into the commander's seat, slamming the armoured door shut hard, getting my map out while resting my rifle across the top of my legs under control of my right hand so if we had to debus quickly it would be ready to use. All this time situation reports were coming in over my earpiece via my radio.

Matt was my 2IC: he was on top cover, my under-slung grenade-launcher (UGL). Benny, a big strong Fijian, was my light machine-gunner (LMG). The back doors slammed shut with their distinctive familiar sound, with both guys in the standing position ready on top cover. We started to move out; Ian as lead vehicle with mine following on at a distance.

Meanwhile the sound of gunfire could be heard over the radio and above the engines of the vehicles. We picked up speed, heading towards the front gate of CIMIC House. The whole compound was alive with movement, a few men fully geared up running towards the Sanger positions. The gate sentry pushed the gates open with some force and some help from an Iraqi policeman. We sped past the two of them and the front Sanger out into the street ahead and picking up speed, motoring past the Pink Palace and police guards on our left.

Ahead the view from inside our vehicles was a straight tarmac road with a central reservation dividing it in two. The reservation was made of concrete, a bollard effect, about 3 feet high and interspersed every few metres by a lamppost, road sign or advertisement of some kind. On our left were buildings two or three storeys high with small alleys and alcoves running along the sides. These were dark and filled with rubbish. Above the first level of shops and cafes were balconies covered with satellite dishes, assorted antennas and washing fluttering in the breeze. Telephone cables ran along the sides of the buildings directly above us, onto the lampposts and out across towards the River Tigris. The river was to our right, its banks grassed in parts and in others simply followed by old iron fencing mounted atop a low-lying wall which ran along in small disjointed sections. This was about waist height in places and continued all the way down towards our intended location of Yellow Three and the bridge.

Immediately within 300 metres of leaving CIMIC we came under heavy fire: you could hear the splatting of the rounds on the side and front of the vehicle. Inside Taff was sweating like a madman, wrestling with the steering wheel and swerving to dodge debris that had been placed in the road ahead. I found myself holding on to the dashboard with one hand and with the other trying to stop Taff's rifle from smacking him in the head as it had bounced out of the rifle rack where it was secured in the cab centre. Straight away in response Benny started firing the LMG in short bursts and every few rounds tracer could be seen flying down the street: it was incoming from all directions. Matt also started to fire his rifle. Empty cases and link were being ejected from both their firing: it started rolling down over the windscreen and falling down into the back of the vehicle. At this point Ian's vehicle broke left, smashing over a narrow rubbled gap in the central reservation, nearly taking off, severely swerving and leaning heavily to the right. For a split second I thought it was surely going to roll. Lee, his very experienced recce driver, somehow kept control of the vehicle, bouncing up and over the obstacle with the top cover guys being thrown about in the hatch like rag dolls.

Taff started shouting: 'Vehicles to our front, they have blocked the fucking road, the bastards.' Straight in front of us were two burning cars blocking the road ahead. We had no option but to smash through them. I shouted up to Matt and Benny: 'Hold on, fucking hold on, NOW!' We smashed our way straight through the vehicles. Now, directly in front of us we could see through the smoke and burning debris that had been deliberately put in our way. Dark figures could just momentarily be spotted running out in the distance to our front and sides, across the road and in and out of the side buildings. Movement was also seen up on the rooftops, with figures running along towards the bridge as well. All this time the sounds of rifle fire and the dull thud of explosions were growing louder as we approached our intended location.

We were now halfway down towards Yellow Three. Ian's vehicle was forward right on the opposite side of the road, smashing everything in its path as it moved at speed. I was over to the rear left on the other side of the road, trying to break the sound barrier in my efforts to keep up with Ian's vehicle. Meanwhile Benny and Matt were firing and shouting down target locations that they could see. The PRRs were simply no use: due to the amount of background noise and the buildings all around us, we couldn't hear each other on them.

Then just out of the corner of my eye, forward left from behind a low-lying wall, I saw a bright yellow flash and something moving towards us at speed and at waist height, moving erratically with a small smoke trail behind it. Within seconds it hit the front left-hand side of the vehicle, just above the main headlight area that was protected by a metal frame. It immediately flew up into the air with its smoke trail left hanging for a few short moments, disappeared out of our sight and exploded behind us. This was our first experience of an RPG warhead being fired directly at us. I shouted back and looked back, and could see that Matt and Benny were still standing up in position. I thought: 'Thank fuck, they're okay, everything is good, we're still mobile.' The sweat was stinging my eyes, my head pounding in my helmet, my heart beating so strongly it felt like it might crack my sternum open. Taff looked to be the same way. His face was red, and with sweat pouring down his forehead, he was shouting and cursing with every word that came out of his mouth while trying to drive and dodge everything in the road.

We were now coming under an even heavier rate of fire directed towards the two vehicles. It seemed to be a continuous wall of fire coming from all directions and from above us. Abruptly the front left-hand right side of the split windscreen cracked in three separate places: three rounds had just smacked into it and a cracking effect, veining, appeared in the panel of armoured glass right in front of me. On seeing this, Taff swerved the wagon across the road onto the same side as Ian. My top cover Benny and Matt were all the while firing, giving target indications to each other and trying to stay in a stable fire position in the top cover hatch. The same was happening with Ian's top cover. Both vehicles then came to a rapid tyre-screeching halt roughly 20 metres apart just under the bridge at the position known as Yellow Three, which was now above us. Nothing was moving above us on the bridge and we were well aware of the possible threat from where we had stopped to debus.

What we could see to our front was the road continuing straight ahead, now with multi-storeyed buildings and alleyways on either side. Slightly rear left of where we had stopped was the Al-Sadr building, literally within metres: a square edifice, light brown in colour, with two floors windowed top and bottom and a balcony facing towards the river. A low white wall surrounded the front of the building, facing the river: within this area it was grassed with a narrow path leading to the main iron-gated entrance.

The road running left past the building itself ran parallel to the road coming down off the bridge, again with buildings on either side and alleyways between them. We could see the burning Snatch vehicle on the road. Smoke was coming off it towards the sky; the smell of burning plastic and rubber filled our nostrils and was horrible to breathe. We could hear rifle fire everywhere and a large contact going on to our left as we approached the vehicle. We dry-fired and manoeuvred towards the burning vehicle. I remember seeing traces of blood on the ground and thinking 'Where the fuck are they?' as we approached the vehicle, 'Where the fuck have they gone?' It was empty, the vehicle was fucking empty. Shit. Just at that moment the ground started spitting bits of gravel up around us and rounds were splatting into the side of the Snatch vehicle, right where we stood. That now-familiar sound scares the shit out of you: the crack and thump of rounds whizzing just past or above you. Within micro-seconds we turned around, fired and manoeuvred back to the cover of the bridge. The remainder of the call sign just let rip with 5.56mm and UGL fire, providing cover for us.

We made it back: we were only about 30 metres away. I got in behind the cover of the wall coming down off the bridge and tried to get my breath back. The air was hot and filled with smoke and my helmet felt like it was trying to crush my head. I changed magazine, my hands shaking, and I remember looking around for a second. Everyone – Matt, Benny, Chris, Ian, Taff and

Lee – was in his own little world of conflict. I popped back up facing down the street, rounds coming in from all directions and landing, impacting all around us. About 80 metres ahead on the corner of a building I saw a figure with an AK firing: I just aimed centre of mass of what I could see, and fired and fired until the figure went down and slumped to the ground. People were engaging targets all over at ground level; there was no time to think. Then suddenly Ray, who was a member of Ian's team, appeared next to me. We were both in the kneeling position at this point. Ray tapped me on the shoulder and pointed to the roof of the Al-Sadr building: there was an RPG man standing up and about to fire directly at us. We both instantly engaged the target with a rapid rate of fire: he fell from the roof to the ground in a heap. Ray's hot empty cases were hitting me in the side of the neck and bouncing off my helmet at the time and the unforgettable smell of cordite filled the air.

We were literally side by side, shoulder to shoulder along that piece of cover, firing and changing magazines from that position by the waist-high embankment, all of us in some variation of the kneeling position. By now the enemy was gaining in numbers and firing from the rooftop, balcony and windows of the building. We were receiving a heavy rate of small-arms fire at this point, mixed with RPG fire which started coming down from the rooftops with some very near misses. Along with that, the odd grenade or two was being lobbed from behind the walled entrance area. We were now within 50 to 60 metres of the building itself but could not stay there as the enemy would very soon outnumber us. They began manoeuvring around to our flanks in small groupings while engaging us, using the alleyways and narrow streets and their street knowledge to their advantage. The enemy that was already using the building as a strongpoint was now starting to co-ordinate their fire with fighters out in the street.

It was necessary to move, and move quickly. We moved directly under the bridge with a little fire and manoeuvre to get ourselves there, using the embankment and the pillar supports of the bridge for firing positions and cover. After a mini re-group all were accounted for and all were okay: what a relief, so far no one was injured. We sorted out our ammunition, redistributing it among ourselves and readying the extra ammunition for the LMG gunners.

All this time rounds were incoming towards us in our new location; all around us was sheer violence and noise. My next target was in one of the side windows of the building. He presented too much of himself trying to empty his magazine towards us and I couldn't miss: I fired at the target; the target went down immediately out of sight. Matt was engaging the area to the main entrance of the building where a few insurgents were using the low-lying wall as a fire position and for cover. Benny was firing in small bursts: tracer could be seen flying towards that location; a burst, then the splat of the rounds impacting at the other end.

By now we knew that Dan's call sign had moved off with the casualty and were having their own street battle moving away from us. After about ten minutes under the bridge the fire started to calm down. Ian and I were kneeling by one of the pillar supports; Ian was trying to get communications with the operations room and Dan for more information. The remainder of the call sign was in all-round defence, while the section's 2ICs were redistributing the ammunition, conducting another head check and enforcing the drinking of water, on which we were all very low. The Snatch Land Rovers were manoeuvred towards our static location: they had been damaged by shrapnel and small-arms munitions with clearly visible strike or impact marks all over the vehicles' bodies and armoured windscreens but they were still roadworthy.

We came up with a plan. Then from about 200 metres to our rear, an RPG with its signature smoke trail and streaking noise came towards us from the corner of a building. Within micro-seconds we looked at each other, then back towards the threat, and then threw ourselves to the ground, at the same time shouting: 'RPG REAR! RPG REAR!' and trying to get around the other side of the concrete pillar. I remember curling up and shutting my eyes, thinking: 'This is going to fucking hurt.' Then in an instant there was a flash of light that I could see through my closed eyes and I felt a wave of heat go over my whole body and an almighty bang in my ears, all simultaneously. When I opened my eyes, my head felt like it had been hit with a sledgehammer. I remember straight away slowly moving my legs, thinking they're okay, they're still there. My right ear was killing me: it felt as if warm water was running out of it, and on top of my right shoulder I could see small specks of blood starting to appear on my shirt.

Within seconds I felt a stinging pain in that area and I can only describe it as a real warmth, plus a slight numbness in my upper arm and hand. I sat up slowly, as did Ian. He was saying something to me, I think, and shaking me at the same time but I couldn't hear anything that he was saying. I just watched his mouth move, which seemed to happen in slow motion. Then sound started to come back to me, muffled at first but becoming more understandable as the minutes went by. Pain, sharp jolts of pain, is the only way I can describe the right side of my body for the next few minutes. Then adrenaline took control as I came around, back into the game, and tried to sort myself out.

Again we both moved behind the pillar and could see what was going on around us. It was around this point in time that the CO's Rover Group joined the battle as they were in the area visiting CIMIC House. Upon arriving at our location they pushed through our position and 100 metres further up the street in their vehicles where they too came under heavy fire and dismounted, taking up fire positions and returning fire.

Back to the plan. Rounds were flying around everywhere: another RPG came over our heads and exploded into the pillar next to us. Time to act. We briefed the men on our plan, then my fire team and I fired and manoeuvred out of the cover around to the other side of the bridge, while Ian's team put down some rapid fire and UGL volley fire towards the building.

We were now lying on the side of the road in the prone position using the high kerb as cover and had gained some elevation from our position, increasing our field of view of the contact area. Straight away we were targets again: rounds started to come in directly from the building opposite us, now directly to our front under 60 metres away, again incoming from the building's windows and rooftop. This time, top right was an AK blasting from that location. I remember aiming at the centre mass and firing towards the muzzle flash of the dark silhouetted figure that was moving about in the window. The firing ceased after I had despatched four to eight rounds at a rapid rate towards the threat.

Benny, Matt and Chris were all firing at the building. Benny's Minimi [a light machine gun] was hammering the window apertures and soon all the glass had been shot out. Movement could be seen in the courtyard below. Forward right from our position there was insurgent movement everywhere. Ejected cases were bouncing off the tarmac road and kerb and going everywhere; we were also getting low on UGL rounds and Matt was now down to his last two. Over the radio the message came that QRF from Camp Abu Naji had been tasked and were en route. I remember Benny saying: 'Well, they better hurry the fuck up, my barrel is starting to overheat and the spare is fucked, everything is fucked, this isn't going to be good, Monty.'

The QRF from Camp Abu Naji had been tasked to come and help both call signs that were in big trouble. Warrior armoured fighting vehicles complete with dismounts were on their way: the Warriors with their big 30mm RARDEN cannon and 7.62mm chain guns were real mobile firepower, fire support. All around us and as far as we could see into the distance were armed figures: some moving around as individuals, others in pairs, and what looked like small fire teams running from cover to cover, building to building, while firing. More enemy fighters were being brought into the area on small buses and pick-up trucks; these were being used to manoeuvre around us and re-supply the fighters. About 60 metres to our rear were some buildings and sure enough, they were soon up there in the windows and on the rooftops, firing from behind us.

We were now up on our knees, engaging to our front, rear and flanks as they were trying to close in on us. The remainder of the multiple managed to extract to an ERV not far from us, while we provided a fire base for them to extract. The message came over the radio that Ian was now in position and we were able to move. We had to move again to rejoin the rest of the multiple to extract: this time we threw smoke out to our front and rear, letting rip with bursts of automatic fire till the smoke built up. Then we moved quickly, a well-practised drill peeling to the left, going back down under to the cover of the bridge.

We got back into the cover, hugging close to one of the thick main concrete pillar supports of the bridge. By this point we were all breathing out of our arses and striving to get our breath back. The warm air surrounding us was uncomfortable to take in and made my chest feel even tighter. It was air filled with the unforgettable smell given off by a smoke grenade and the reek of cordite was heavy too: our nostrils burned with every deep breath that we took trying to get air into our lungs and calm the body down.

Once again, we were straight into all-round defence and a quick head check, then in our pairs a quick magazine change. And over all the noise that was going on around us, we could hear the familiar sound of the Warriors making their way towards us. The Warrior armoured fighting vehicle has a very distinctive engine sound, and mixed in with the clattering of the heavy track going over the rollers on the tarmac road as they powered their way along in our direction, it was the best sound in the world just then. I remember we all looked at each other for an instant and all of us let a little smile appear on our now tired, dirty faces. A micro-second of relief, a moment of calm, thinking: 'Thank fuck for that, armour support at last.'

Then it was back to reality with a loud thud. An RPG went winging past us into the wall on the other side of the road and exploded on impact, sending chunks of wall flying up into the air and across the road in all directions with a blackened scorch mark being left at the centre of impact. We needed to move now and rejoin the rest of the multiple. Ian and his half of the multiple were now securing our ERV and the two Snatch vehicles; we had to make our way to them.

Information started to come in over the radio from all the call signs on the ground. The CO's Rover Group were now on the ground and forward of us dismounted. They were involved in a large contact, taking incoming fire from all directions and RPG fire from the direction of the bridge area itself. We also heard they had a casualty: Kevin, one of the CO's Rover Group dismounts, had just received a single gunshot wound with the entry point in the shoulder and the exit point in the side of his neck.

While dealing with that situation they were continuously under heavy gunfire and grenade attack from all directions. Kevin received basic first aid on the ground in cover and was patched up so he could be moved. He was put into one of the Rover Group's vehicles and extracted back to the medical reception centre at Camp Abu Naji. Meanwhile we started to fire and manoeuvre down the street, keeping to the left side and moving along with the River Tigris behind us, peeling to the right, rifle up, moving in short bounds and down straight into the kneeling position, observing to the front, left and right and scanning the windows and rooftops. It sounded as though the whole city was in chaos: gunfire, both single shot and burst fire, could be heard sporadically, plus a mixture of explosive sounds and dull detonations among the small-arms fire. The best sound of all was that of the 30mm RARDEN cannon firing from the Warriors, interrupting the chaos.

We were approaching a prominent junction that we had to cross and stopped short in a small garden to overlook the obstacle. I put Benny out with the Minimi to cover us with Chris as his cover man facing to the rear. Matt and I were about to go for it and cross the junction when to our front within about 20 metres two armed figures dressed in dark clothing with one of them wearing a tan chest rig suddenly appeared on the corner. They were as shocked as we were! They were both in the standing position: Matt and I fired at the same time. Fucking rapid fire – operating the trigger as fast as I could – and the targets went down where they stood. We waited and watched to see if any more were about to appear in the same area: it felt like a lifetime, with sweat dripping down the inside of my helmet, into my eyes and down the sides of my mouth. I was constantly wiping my face with my face veil, which was already heavily soaked and had little effect, but this all happened in a matter of seconds. We looked at each other: 'It's clear, let's get going', and 'Move now!'

We were up and moving rapidly across the junction, hard-targeting. Once across we got down back into the kneeling position, scanning all around us. At this point I couldn't hear much over the radio. We knew the CO's Rover Group had a casualty; the Warrior call signs were dealing with the extraction by providing protection and firepower; and Dan's call sign had been or was about to be extracted and their casualty was conscious and patched up with a shrapnel wound to the lower leg.

Once Matt and I were in position we signalled for Benny and Chris to come over. We were now in another garden with a low-lying wall that surrounded a very small single-storey building. Benny faced the direction of the building covering the doorway and single window; Chris covered the rear of our position (the direction we had just come from); and Matt and I were peering over the wall looking in the direction we were about to take. We now could see the ERV which was to our front about 200 metres away, and we could see the Snatch vehicles. We were really low on ammunition by this point: Benny was down to his last fifty rounds or so of the belted ammunition on the Minimi; we had no UGL rounds left; no smoke grenades left, only my phosphorus grenade; and only a magazine or two left of our rifle ammunition.

We took a couple of minutes to sort ourselves out and calm down, a couple of gulps of warm water from our camelbacks was downed, and at the same time we made sure that the magazine we now had fitted on our weapon was full and ready to use. With that, we were ready for the final push down to the ERV. The battle seemed like it was moving away from us. Dark plumes of smoke could be seen going up all over the late afternoon sky from small fires of burning tyres and vehicles. Sporadic gunfire and the odd explosion could also be heard.

We moved as we had been moving all afternoon: short, sharp bounds from cover to cover; no need to scream and shout, it was now a slick drill. As we approached the ERV I could see Ian waving us in. Lee stood by one of the Snatch vehicles; the remainder of his multiple was in all-round defence observing out. I moved in straight on Ian. I could hear him now. I briefed him on our turn of events and he started to back-brief us on what had just happened, who was doing what, and the extraction plan. We were all to move back to Camp Abu Naji as a convoy with the Snatch vehicles in the middle and the Warriors to front and rear as protection.

Then suddenly Taff appeared. He came over to me with a piece of RPG shrapnel sticking out of the front centre of his ECBA plate worn in the middle of his chest. I said: 'Fucking hell, you are one lucky Welshman. That's fucking mad, mate – you tried to get it out?' He replied: 'Yeah, I can't bloody move it and I don't have my Gerber on me. It's bloody stuck in there and sharp and really pissing me off.' It was good just to see him and we just laughed at each other for a moment. We were all in. 'Let's mount up' came from Ian, and with that we mounted up into the Snatch vehicles which were now looking a bit battered. I climbed into my seat, my legs and everything else aching and damp from sweat; it all seemed such an effort just to do the most minor task. I slammed the door shut. We were all chinned [exhausted] by now as we had been on the go for over four hours. Taff was already in and waiting for Benny and Chris to slam the back doors shut, and with that we moved off.

It was all quiet in our vehicle as we drove in convoy back to Camp Abu Naji, still with the sounds of sporadic firing going on in the distance that could be heard over the noise of the vehicle engines. As we drove through the streets they seemed empty of people as if they were all still in hiding.

It just seemed unending: debris of rubble, broken glass and pieces of metal scattered all over the place, mingled with rubbish being blown around by the wind. The few burning vehicles left in the road were just smashed out of the way with ease by the Warriors so we could continue on our given route. After about fifteen minutes of travelling and, on a personal note, with a constant feeling of uncertainty and unease as to whether they might strike again and from where, the small convoy arrived at Camp Abu Naji.

We went straight in through the front gate and directly to the area by the unloading bay where we came to a rapid stop in a cloud of dirt churned up by the vehicles, covering both them and the top cover in a fine, thin layer of brown dust, much to the top cover's displeasure who then swore and cursed at the drivers. We all got out and made our way to the front of Ian's Snatch vehicle with our weapons. I remember taking my helmet off: my head was pounding, my hair matted to my scalp with sweat. To feel that slight breeze on my head; well, the relief of removing my helmet just seemed so surreal and calming to me. We all looked the same: sweaty and dirty with bloodshot eyes. There was a moment or two of silence, then the smiles started to reappear and with that a little laughter and piss-taking of each other.

I moved over to the unloading bay and supervised the unloading of the weapons. One of the guys had a stoppage on his rifle and could not clear it. I remember using the three-piece cleaning rod and clearing the obstruction from the barrel, thinking: 'How long has his rifle been like this? Has he been running about all afternoon in the shit unable to fire his weapon or not?' That's what got to me. Then once that was done we had to get a re-supply of various ammunition. Equally important was water for the men, sorting out our radios and preparing the vehicles for the move back to CIMIC House as they needed to be refuelled. All this happened concurrently: every man had a task and got on with it. Ian and I went to give a back-brief on the events of the day for our call sign and were then briefed for the move back into the city. We were not away for long and came back to the men. Now the kit and equipment was ready and redistributed around the multiple.

By now we had all calmed down. Our bodies started to go a little into shock, with dehydration starting to kick in together with awareness of all the little cuts and knocks on knees and hands that went unnoticed until now but started to throb or sting once the adrenaline had gone. Corky, our medic, came to check the boys over and he took a look at my ear and shoulder. The ear hurt more than the shoulder at this point, but my shoulder had what looked to be a few small, dark, thick scratch marks which were very small splinters of fine metal from the RPG. A quick clean with a sterile wipe and crack on: nothing to worry about and I thought I was lucky enough. Ian was ragging me, mucking about, saying: 'Monty didn't move quick enough, that's why you got fragged because you're a blind old git and didn't see it coming.' Then Corky tried to join in the attempted ragging while he was sorting me out, and all the time Ian was comparing himself to GI Joe, having saved my arse.

By now everyone had taken some water down their necks and we managed to get some chocolate as well. That went down well, together with the good old favourite, Haribo (the food of the gods). Then the order came: we were to move out in twenty minutes; to load up our weapons and mount back up into the Snatch vehicles and get into the order of march for the route back through the city and into CIMIC House. We formed up with a lead Warrior followed by Ian's vehicle, then mine, and another Warrior to the rear. We checked the communications between all the vehicles in the packet, then it was time to roll out. The top cover were up and ready: Benny facing forward with the LMG and Matt to the rear with his rifle. It was now late in the afternoon; the wind had picked up and was starting to blow sand and the odd bit of foliage around our vehicles.

The sun was starting to go down as the small convoy started to move out slowly in its order of march. The sound of the Warriors with their powerful engines roaring into life, surging forward with thick black smoke coming from the top of the vehicles' exhausts, the vehicles suddenly stopping and then surging forward again, the grinding of the track over the rollers and rear idlers, metal on metal: you never forget that sound, ever.

We passed through and out of the camp via the front gate and past the elevated Sanger position which was covered in a massive desert camouflage net and the battalion sign. Taff and I waved to the gate sentry from inside the vehicle. Benny and Matt were heard shouting something to him and laughter came from both the sentry and the top cover. Inside, I was orientating my map to our direction of travel. We were going the long way round into the city to avoid the areas where we had just been in contact, out of Camp Abu Naji down the long, dusty, potholed road and under what I can only describe as a double archway that in parts must have been damaged by some heavy ordnance in the past but was still standing. We then turned left onto a tarmacked road and the convoy increased its speed. Coming into the outskirts of Al Amarah we just went for it: buildings either side of us all along the route similar to the ones earlier that afternoon just passed us by in a blur.

Past the hospital and over a bridge, we were now on a straight road with the convoy making good time. To our left was a high kerb running along the entire length of the road: it was filled with earth and slightly grassed, and this separated the road into two lanes for traffic. Benny and Matt were letting me and Taff know what activity was going on to our flanks when suddenly there was an almighty explosion: our first IED. The vehicle rocked from side to side, while simultaneously a cloud of dust and dirt engulfed everything. We were still moving for a split second, with Taff shouting: 'Fucking bastards, fucking shits!' He slammed on the brakes at the same time and as he did so the vehicle lurched forward and screeched to a halt, rocking backwards and forwards for a moment or two. I looked back and could see that Matt and Benny were still standing. I was grabbing Benny's trouser leg and leaning back, shouting: 'You two okay? You two okay?' I was about to get out but then they both started shouting down to the cab and swearing like mad. They were okay, thank God; He must have been on watch that day, over all of us.

As the dust and debris started to clear we could see the front of our vehicle and the Warrior ahead with the bright red glow of its rear lights piercing the settling dust cloud. Luckily for us the armoured fighting vehicle had taken the main blast of the explosion to its side panels and was largely untouched, the tracks were still in place and functional, but a crater had been left on the side of the kerb and road. Ian's vehicle was okay; he opened his side door and signalled to me that the men were all unharmed as well. Straight away the lead Warrior signalled and moved off like a bat out of hell, followed by Ian's vehicle, then mine, and finally the last Warrior in file.

Sheer volume of noise, with the vehicles' engines screaming and black smoke coming from the exhausts of all four on acceleration, followed us down to a roundabout where we put in a hard right turn, the Warriors' tracks churning up the tarmac and leaving black marks on the road. Accelerating out of the tight turn, we were then straight down the road towards CIMIC House. When we arrived and got out of the vehicles everyone was looking like shit: dirty and weary, still sweating from every part of the body. But still a little smile and laughter came to our faces within minutes of everyone seeing each other and realizing that we were all more or less okay. Then it was straight into the unloading bay to unload. Now at last we could take off our helmets and body armour, webbing or chest rigs: such relief on everyone's faces with the sound of Velcro being ripped apart to get the body armour off and the plates hitting the floor where we stood.

Everyone who smoked seemed to light up straight away or was trying to find their fags or lighter, and at this point we all sat down in a big group and started chatting and joking about. Chris went to the side of the cookhouse and came back with a box of bottled water for the multiple. This then became the drill we adopted after every contact or incident: the getting together and general discussion. It does help up to a certain point; everyone talking about what they have just done and experienced and how they are feeling now. Also useful for getting facts, such as the information to be collated for a report or back-brief. On this occasion Ian and I went up to the operations room to back-brief the company commander once we had all the necessary information.

Within a few hours everyone was in their sleeping bag and trying to get their heads down. For me, I remember lying there unable to sleep for a very long time: to this day that night is still very vivid in my memory. My body was aching all over. My head was pounding: to me, the noise inside my head was louder than the noise outside my body. Every time I closed my eyes the events of the day would reappear in my mind in flashes of colour like a detailed replay on a DVD, or just quick flashes of random images from the events. My mind was racing, asking so many questions. What if this or that? What if an incident had turned out a lot worse than the actual outcome? Could I have prevented it or changed it? Why did it happen to us or to me?

Should I have fired and engaged that target, that threat? The answer for me is always an unequivocal YES.

That was just one day, our first day in contact. What a day it was, and there were to be many more days and nights like that to come, street-fighting in Al Amarah in 2004. On that first day we had all been baptized into a very special brotherhood: that of the modern fighting soldier. And some of us would continue to soldier and fight on in another time and place in the not-too-distant future.

Chapter Two

Afghanistan: Pre-tour Notes and Thoughts
Operation HERRICK 8/9

Afghanistan is a country with a very turbulent and violent past filled with conflict and oppression through the ages until the present day. There is an old saying that I believe comes from an old war film, which says you can only tell the decade in Afghanistan by the colour of the uniform the soldier is wearing. I had seen a few of the war films made about Afghanistan, from the days of the British Empire through to the Soviet invasion and occupation, and that was the image of the country and her peoples that I had imprinted in my mind: a large country of hardship with an inhospitable climate to match. The spectacular and arduous terrain was also matched by its people, who were just as hard as the land and who needed to be in order to survive. Recalling my experiences in Iraq, I had learnt some basic lessons and after a bit of thought the old classic question from the Infantry Battle School came into my head: 'If you could do that again, what would you do to improve on it and why?' These were my thoughts from my earlier experiences:

- Never, ever underestimate your enemy. It only takes one well-placed or lucky round or device from him and it's Endex for you.
- I am not the best: there is always someone better. Regarding counter-sniping, just consider and be very aware that there may be a well-trained and experienced enemy sniper out there operating on his home ground with a good solid understanding of his role. He will have local knowledge of the terrain, may be acting with the support of the people, understands the environmental effects of seasonal change for his area of operations, and knows what limitations may affect him. Even a combat-experienced insurgent marksman with just a basic optical sight crudely attached to

his weapon system is a real threat, especially at closer ranges or operating within the built-up areas of villages and towns.

- I am in his country, his homeland, and not everyone wants us there. To some in that country, I am a foreign invader. This is a very good tool for his or her motivation and ideology which can be a strong unifying force for some of the native people in their fight against us.

- Every contact or kinetic engagement with the enemy is a lesson learned about them, so learn from it. Seize the initiative: you act first. If the enemy is pre-seen, take the fight to him within your role. Take time to study the ground and environment where possible, and make good use of any available assets.

- Sniping from helicopters (or heli-sniping) and the theory behind it both sound good. But in my limited experience in this area – two times up and both times getting into the shit – it could have potentially been fatal for the crew, for us, and resulted in the loss of the airframe. In reality, once contact is made with the enemy you have become a bullet and RPG magnet; a priority soft target for him. When the enemy is in force, mobile and with a concentration of firepower directed towards you, you are extremely vulnerable in the air. Again from my own limited experience, only the sheer skill of the pilot plus the agility and speed of the Lynx helicopter saved the day on both occasions.

- Finally, a single-shot bolt-action rifle with a magazine of ten rounds is not always the best option, especially when engaging with a very dangerous enemy at ground level in the confines of a built-up area. Do not get caught out: your situation can change very rapidly to one where the circumstances are no longer in your favour. Always try to imagine the worst-case scenario and calculate your actions to counter this when making your plan. Working together, the small grouping of a reconnaissance platoon and snipers combined can observe and report, fight and protect itself, combining very similar skills to achieve the operation aim.

A lot of snipers and their No 2s prefer to carry a pistol as their secondary means of self-defence as it is small, compact and easy to bear. The use of a pistol could depend on the task in hand, along with the threat to the sniper pair at the time. A semi-automatic rifle is generally preferred, giving greater firepower and increased range over that of a pistol. For me, my rifle SA80 A2 was either in my hands while moving from position to position or on my back, and armed with twelve magazines of 5.56mm ready to use. An example for its use might be entering a compound and clearing the location in order to gain a position of elevation, to gain access to the rooftop and be able to observe over a target area or to engage an enemy. The biggest threat to a sniper is always another sniper: they will use all available means to take you out.

Depending on the type of enemy and their threat capabilities, as soon as you are out of your patrol base on task, either on a foot patrol or mobile, you are being watched from the moment you leave your base till you return. The enemy will very quickly work out your command structure and who does what in the patrol. They may also be aware in some cases what our various pieces of equipment are used for and why we use them: nothing goes unnoticed. They are able to recognize your weapon systems and capabilities, observe your skills and drills, and may well test you with an engagement of some kind to gauge how you react to that threat or situation. Always when prioritizing a threat, a soldier with a scoped-up, possibly camouflaged rifle is the priority target and must be eliminated, given the opportunity.

My kit and equipment had greatly improved since my past tours, and for this tour one massive upgrade was the new sniper rifle now being fielded to the British army: the L115 (A3), this being the newest, most improved .338. The rifle is heavier and longer than the L96, carries a bigger punch at the target end, and can also be fitted with a suppressor for engaging a target from a concealed position with minimal signature at the firer's end. The rifle's range is greatly increased and it is fitted with a much more powerful scope, allowing a sniper at the greater ranges to more easily identify and engage a target. I also had better body armour with larger front and rear plates which not only offered better protection but was more comfortable for prolonged wear, and a new improved helmet again offering superior protection.

I carefully prepared my kit and equipment accordingly to deploy: everything was packed, repacked and double-checked. I also ensured it was waterproofed as well: doing this helps to keep the sand and bugs out of your kit, wherever you may end up. Green tape and spray paint are the sniper's allies: a DIY kit enabling him to enhance his equipment to his own taste for the environment in which he is about to operate. The use of various netting and foreign camouflage patterns is also a favourite, much to the annoyance of the powers above. It is always the reconnaissance and sniper platoons of a regiment who are masters of winding up the CSM or RSM regarding kit and equipment and how it is worn by the men!

Prior to the tour of Afghanistan I had been a sniper instructor for over eight years and had a few operational tours under my belt, both as a sniper and a member of the reconnaissance platoon. I also served both in the brigade surveillance and COP platoons (in which enduring patience was always required). I was lucky in the fact that I joined back up with the reconnaissance platoon for the tour of Afghanistan and that as the platoon was fully structured and manned, I was not needed as a platoon sergeant. My task was to deploy with the platoon as a sniper with my new improved .338 and also give advice if and when asked on matters relating to sniping. This might be with regard to employment, capabilities or simply how to use snipers to maximum effect, both on routine taskings and on more specific planned operations in support of other friendly forces' call signs out on the ground in our area of responsibility. Just as important to me was the fact that I had three fully trained and motivated snipers from within the reconnaissance platoon for on-the-job continuation training, so to speak.

The idea that the boss Capt B (MC) and I came up with for the snipers within the platoon was a simple but effective one, aimed at them all gaining some experience and knowledge on operations deployed as snipers and putting this to good use in the future. They would rotate, spending several weeks at a time away from the platoon with me in a patrol base, and we would move from location to location as needed, concentrating on the sole task of sniping and counter-sniping later on in the tour. The theory sounded good at the time: a cracking plan for the next generation of up-and-coming snipers to put into practice what they had trained so hard for, principally to get out on the ground with their rifles.

All three of the guys were lance corporals from within the platoon. Ben came out with me first: he had completed his basic sniper course at Brecon and gained a distinction upon completion. He was to spend most of the time with me due to Carl and Matt being called away on operations with the remainder of the platoon, either as drivers or as gunners in vehicles such as the Scimitar. They too were fully qualified and each carried an L96, but only because they had not yet done the conversion course to the new L115 (A3) .338 rifle.

Ben and I each carried a .338. We were a sniper pair, Ben and I, and some might ask: 'Why two .338s?' Again speaking from experience, two is better than one: we could engage different targets at the same time and it gave us good stopping power at greater ranges. It also had good penetrating power through any equipment worn by an insurgent, including the basic protective vest. The increase in magnification power of the rifle scope was up to x25, excellent for positive identification of a target and for observation.

Finally, a very important skill that we practised on the ranges including Bastion ranges in theatre was co-ordination shooting: both snipers, having identified a priority target or targets, would simultaneously engage the threat and neutralize it. This skill can also be applied by using not just one or two sniper pairs but multiple pairs deployed on the ground up to platoon strength with devastating results if carried out correctly: firstly by physically destroying or neutralizing the enemy, quickly and silently; and secondly mentally, by affecting their morale.

Where all such pairs are in a position that affords them good eyes onto a location and all are able to positively identify the targets on the ground, once these have been prioritized and indexed by the platoon commander the sniper pairs are given their marks. Then with all the pairs simultaneously engaging their respective targets at the same time, on command from the platoon commander on the ground the threat will be neutralized instantly. Hopefully all targets will drop at the same time but it may be necessary to finish off the area with a fire mission, e.g. by dropping mortars or artillery onto the position. Also destroying any kit and equipment including ammunition, food, soft-skinned vehicles, light armour or crew-served weapons will remove them from the enemy and help to deny him the future use of that position.

Chapter Three

Nad-e Ali, Patrol Base Argyll

Part I

It was now mid-November and the very tail end of yet another long, hot, sweaty summer in Afghanistan was coming to a close. The dry and dusty days with their blinding natural sunlight and warm predictable summer breezes seemed to start diminishing very quickly. Those long days began to merge into the shorter, colder days of the early winter with less natural light and, more importantly to me and Ben, a now much more unpredictable wind that had increased in strength and changed in its direction of travel.

This was much more apparent during the hours of first and last light when the wind strength and direction must be calculated regularly throughout our hours of observation. Particularly important at these times, as this was when we had engagements with the insurgents on an almost daily basis by now. Large, white, puffy clouds filled with either rain or snow started to fill the early morning sky as they travelled high above us. Now the shorter early evenings suddenly moved into the complete darkness of night and became bitterly cold, with many hours of darkness until the sun rose once more at the coming of a crisp new day.

Above the cloud cover and high into the sky beyond were a few vapour trails, stretching out and disappearing over the horizon. These were produced from the jet engines of our coalition war planes as they flew their daily sorties – Fast Air, as we called it – either to gain intelligence on the ground by taking air photographs or providing fire support to troops in contact on the ground, unleashing their ordnance against the insurgents with pinpoint accuracy and neutralizing the threat. In the late afternoon the slight breeze would pick up a little just before last light, just when we may be trying to light the Hexi blocks to cook our boil-in-the-bag evening meal and make a brew. The temperature would start to drop, just enough to make us put on a softie jacket or buffalo jacket to ward off the chill.

The main position that Ben and I were to use for conducting observation over our location and that for the MSG grouping already sited up there was located on a prominent rooftop within the patrol base, a building which had been an old school house at some point in the past. It was a single-storey, well-constructed, solid building with some sizeable rooms that must have been the old classrooms: one of the rooms that the mortar lads were living in still had a large blackboard fixed to the wall with some chalk drawings and writing still just visible on it. Those must have been drawn by some local Afghan children a long time ago. All the windows around the building were large and double-sandbagged, completely blocking up the whole window frame from the bottom of the wide window ledges to the ceiling. They were also boarded up for protection from the base of the windowsill almost to the top, sometimes concealing the sandbags behind, just with a little gap allowing some natural light and air into the room where the lads were living. They each had a very tight confined space with just the basics: a camp bed with sleeping bag and a dome mosquito net over that, creating a small green tent effect where they slept and kept all their fighting and personal equipment.

The room directly opposite was another large one occupied by the Police Mentoring Team. I remember it was always very dark, with remnants of burnt-down candles on the tables, the windowsills and in the middle of the floor. To the right of that were a few large rooms evenly spaced along a high-ceilinged corridor with large high archways opening onto a courtyard area of mud and dirt with a large radio mast in one corner; this was where the Afghanistan National Army lived and worked. To get up onto the roof we had to climb an improvised wooden ladder. It was only about 14 feet or so in height. But laden with all my kit and both rifles, one in my hand and the other on my back, trying to get up there especially in the dark and with me having the agility of a donkey, I thought this ladder might get me before the Taliban did.

The rooftop was sandbagged and built into a defensive firing platform in which we all could adopt the kneeling or sitting position for comfortable firing, and in the corners adopt the standing firing position if needed. This allowed us limited movement atop the old school house. In the far left corner were three general-purpose machine guns mounted in the sustained fire role. Sited more or less in the middle of the rooftop was a Javelin missile system, and to the right of that in the right-hand corner was the 40mm grenade machine gun. All these weapons were manned by men of the Royal Marines from 42 Commando, J Company MSG grouping. This was their AOR and patrol base in which they conducted operations and fought, both by day and night. For some of the Royal Marines this was their third tour; they were a very experienced and professional small grouping.

The mortar support in this location was from 2 PWRR and they were the theatre reserve battalion, stationed in Cyprus at the time. They manned and fired the 81mm mortars in their dug-out sandbagged pits, which would very soon become constantly filled with water and mud when the rains finally came. The whole area around and inside the patrol base would soon become a constant mud bath for a month or so: we had to fight from it and live in it.

On the rooftop Ben and I had our first chance to take a look at the surrounding area and ground around our location in detail. We were a pair, both with our brand-new .338s, and finally out on the ground eager to crack on with the task in hand. The many hours spent up on the roof conducting observation and data-collecting were never wasted: we watched the pattern of life, the daily routine of the civilian population around us, figuring out who lived in which compounds, which local farmers worked in which fields and so on. Then there were the locals from outside our area. We meticulously observed their routines and habits throughout the day as they moved in and out, also paying attention to what they were wearing and with whom they were associating. Just as important was to gain information on local use of motorbikes and other road vehicles as they transited through our area. What were they transporting? And to what destination?

Animals including cattle, stray dogs, wildcats, birds and any other creature large enough to notice may be very useful as sometimes their reactions can be judged to alert you to a possible threat. It can be a good combat indicator, so sometimes watching animals where they feed, water and rest is important. A sudden change in an animal's behaviour, such as making noise late at night or movement from the herd or pack when you know they normally lie up, may mean that something or someone has startled them.

We would also take the chance to observe light conditions throughout the day, particularly in places with trapped shadow, and to study these in great detail using the Leupold spotting scope. As the sun moves across the sky during the course of the day, it may improve the detail or clarity of features on the ground that are of interest, giving you an idea of their possible practical use by the insurgents as a firing position or point of observation onto your location. Therefore should the insurgents try to use it against you in the future, it will already have been ranged and plotted onto your range card for quick reference.

A vital core skill of the sniper is his ability to be able to judge the wind correctly, i.e. its direction and speed, and to understand the effects that the wind may have on a round when firing, especially at greater ranges. In our location the wind was stronger in the afternoons than the mornings and the direction more or less stayed the same, which was a bonus for me and Ben.

Our first task was to identify known enemy firing positions from previous contacts and to then range them accurately so we would have an exact range to that location should they try to use the same position again. These would be plotted down on our range card for future reference if needed. Such areas of interest would also be observed to see if any of their approach or exit routes could be engaged from our position up on the rooftop.

The compounds used by the insurgents were easy enough to locate and identify as the previous firing positions in and around them were peppered with the splat marks of various munitions, caused by return fire from windows, doorways, any corner of a solid structure, or even the drainage ditches which usually surround these locations. Any holes or cracks in the walls large enough to fit a muzzle through and fire from were noted. By this time of year the foliage had started to die off at ground level with just a few of the larger trees and prominent bushes retaining their leaves. This revealed previously unseen ground, hidden by the greenery during the long summer months: small embankments, ditches, crossing points across some of the irrigation ditches that still had water in them and gaps in-between some of the compounds and derelict buildings were now open to our field of view.

Approaching midday on yet another day of observation and the normal pattern of early-morning local activity had been almost non-existent: today it was too quiet. We had both been on the rooftop in our normal observation position for a fair few hours and visibility was good out to a couple of kilometres in all directions. It was usually at this time of day that the insurgents would attack.

Contact against the insurgents would usually begin with two or three of them opening up and erratically engaging our location with single shot, then within seconds fire was coming from several locations: a mixture of burst and single shot, simultaneously, like springing an ambush. There was the now-familiar sound as the rounds travelled through the air: zipping, snapping and cutting through above and around us, some rounds hitting the sandbags to our front or impacting into the wall just below. Then followed the repetitive thud of either our or the insurgent mortars firing either way: rounds coming one after the other in quick succession. As for the incoming insurgent mortar or rocket fire, well they were just becoming more and more accurate as the days went by, now being able to range in on us.

The insurgent mortars were landing literally just within our compound walls. A couple of dull explosions, and soil and debris were being thrown up in front of our location in plumes of brown dirt up to about 20 or 30 metres into the air and then being carried away by the wind, dispersed back down to the ground as quickly as they had appeared. The rooftop was a hive of activity with the gun controller co-ordinating the SF guns, issuing fire control orders at the top of his voice to the GPMG crews who were rapid-firing. As belts of ammunition were being fed into the guns, the ejected cases and link littering the floor piled up underneath the firing weapons. All three were spitting out 7.62mm in long deliberate bursts, with a little smoke and flash flame coming out of the front of the flash eliminators of each gun as the rounds left the barrel. Gun oil was being sprayed in the working mechanism of the guns to keep them firing. The ground underneath the 40mm GMG was filling rapidly with large ejected cases and chunky link as the gunner fired away, putting down fire into an old firing position that was being re-used by the insurgents.

At this point in time there was a position out on the ground over to our right flank that Ben and I had been particularly observing. We had had our suspicions and doubts about it for more than a week and I remember that I was always drawn to it during an incident: to me it was a position that looked like one of the oldest tricks in the book. We put in hours and hours of observation through our optics and spotting scope towards this location and its immediate surroundings for possible entry and exit routes but we needed proof to PID that it was in use by the insurgents. To me this was a classic

Second World War position for an enemy Forward Observation Officer or Mortar Fire Controller to make use of.

We had regularly ranged it using the laser range-finder to an average reading of about 707 metres away from us, and it was slightly in front of a compound that added to the backdrop of the position on our right. It was a little elevated from its surroundings, and built up just enough to give it a perfect field of view of our location, the compound and surrounding area. The aperture that I was watching was too perfect for me: nature is not that perfect, it had to be man-made. It was very dark and appeared to be a near-perfect circle burrowing back into the position. Yet the aperture had moved ever so slightly up and down over the past week or so. I was quite positive about this: every time mortars or rockets were fired at our location, it seemed to be slightly off centre from its previous position when I measured it last.

The aperture seemed to be about the size of a clenched fist or smaller and on several occasions while observing from different angles during fighting, we had noticed slight movement within towards the rear using the Leupold spotting scope on its maximum setting and adjusting the fine focus. One of the golden rules of concealment had been broken: possibly with no backdrop behind the enemy's observation position, there were a few split seconds of light revealed within the darkness. Perhaps the lens on some form of optical instrument being used by the insurgent was given away by the sun's reflection and caught just for a moment. This was sufficient confirmation: the aperture was man-made rather than part of the environment.

Just one single thin layer of a face veil or other type of netting to cover the front of the optic or even the making of a shroud could have prevented this mistake by the insurgent observer. By now I could be certain that this was an enemy observation post manned by a trained and experienced mortar fire controller of sorts. The mortars had been getting ever closer to our location and it was only a matter of time before they would be landing in and around the position. Suddenly there was a lull in the enemy fire, probably for corrections to be worked out and sent to their firing line. Now the whole roof was bustling with activity, ammunition tins being brought up the ladder to replenish the guns, plus spare barrels so the barrels on the guns could be changed over in order to prevent them overheating. Big bottles of drinking water were being thrown up to us as well.

The Royal Marines on the roof were constantly changing their fire positions along the cover of the sandbagged wall, and on the corners the sandbags were built up a little higher like a small Sanger at each end. Some Marines were adopting the standing position to fire from these corner positions but at the same time still keeping a low profile. Each time new insurgent fire positions were located, fire control orders could be heard being shouted out across the rooftop. Empty cases ejected from the rifles and GPMGs were strewn everywhere, mingled with the sand and link on the floor: when manoeuvring about trying to improve your firing position and adopt a new sitting or kneeling stance you first had to quickly clear the ground with your hand or foot.

Just as we were about to move to a new firing position on the right side of the roof, Ben had PID another target coming into his field of view and gave a target indication to me. The insurgent was at just over 500 metres and our view of him was full-front and moving, making for a low-lying wall and coming head on. In an instant, Ben switched from his binoculars to the laser range-finder and ranged it at just over 510 metres which was good enough. There was no time to waste for either of us; every second counted. 'Ben, you crack on with that, I'll deal with the other. Happy? Let's crack on.'

It was time to act and act quickly. We both moved to our respective fire positions. I cleared the ground with my hands of empty cases and loose link from the 40mm grenade-launcher and shifted a plastic bottle that was half-filled with urine to the side. Sometimes you just have to relieve yourself in situ because there is no other option, and a used plastic water bottle does the job. Or just piss on the floor if you are not going to kneel or sit to fire. I slowly peered over the sandbagged wall and watched the target briefly through my binoculars, resting them on the sandbags to provide a stable platform. I then reconfirmed the target and the range to it, changing to the laser range-finder and trying to keep as still as possible for a final accurate reading. Again it read 707 metres.

The wind direction and strength had hardly changed. However, after a few moments' thought and observation to be sure of my corrections, I checked my elevation drum and deflection drum: everything was correct. I adopted a variation of the kneeling position, forcing my knee pads into the hard dirt, opening my legs as wide as possible to lower my position and provide stability, rested my arse on the heels of my boots and remained hunched down as low as possible. I removed my bipod legs from the front of my rifle, removed my bean bag from my pouch, placed it forward on the edge of the sandbag and patted it down. I placed the front part of the rifle stock on the bean bag and with the weight of my .338 it sank in slightly, cushioning it from the solid weathered sandbag. All this time around me there was firing and shouting: a deafeningly loud noise with the dull thud of the 40mm GMG firing in the background. The familiar smell of cordite hung in the air, filling my nostrils. Above us were two Apache helicopters that were now in the area and involved in the battle below them.

I was concentrating so hard on what I was about to do, trying to block out everything that was going on around me, and focused so intently on the task in hand that the right side of my head felt like it was physically throbbing. This was it: place a well-aimed round into the centre of that possible observation post aperture, that is what I needed to do. The settings on my rifle scope were all correct; now I must concentrate on building up my position. The rifle butt was firmly in my right shoulder, slightly forced into my body armour, under the control of my right hand which was gripping the rifle firmly with my right index finger running just above the plastic stock over the trigger. Sweat was streaming down the front and sides of my face from my forehead and dripping down onto the cheek piece of my rifle, then down the right side of my neck.

My right cheek was resting on the adjustable cheek piece and with that my left eye closed and my right eye opened wide, looking through the scope to ensure correct alignment of the crosshair onto the target. With my left hand I made the final adjustments on the zoom control and focus, bringing the picture in and out of focus, then back into focus until I was happy that I finally had the clearest picture I could get. With a final slight tweak on the parallax adjuster (in short, like a very superfine focus), I was happy with my sight picture.

That was as perfect as it was going to get. The OP aperture, a small dark circle that had gripped my attention for so long, was now in my scope picture at last. The centre of my crosshairs was the centre of my world. All the time I was trying to control my breathing, trying to slow it down to slow my heart rate, calm fucking down, calm down so I don't feel the thud or pulse of my heartbeat in my arms or upper torso. The added restrictions of my kit and equipment increased the stress on my body, hardly helped by adopting an unnatural body position. My right thumb pushed the safety catch forward to the fire position, the tip of my right index finger then rested lightly on the bottom of the trigger. I breathed so slowly and deliberately, focusing on the aperture and the centre of my crosshairs as they moved ever so slightly up and down with each considered shallow breath. On the next up and down movement, my crosshairs were on the centre of the target, the aperture. As this happened, my trigger finger increased the pressure very lightly, pulling the trigger slowly to the rear, then there was a small click as the hammer was released, came forward and struck the firing pin which in turn struck the primer and suddenly ... BANG!

Instantaneously a small explosion occurred within the chamber separating the bullet from its casing, the round then travelling, turning, following the rifling along the length of the barrel at great velocity and power, and finally released on the way to its selected target. Immediately the recoil from the action of operating the trigger was felt in my shoulder. I kept the trigger squeezed in so as not to disturb my position and held it for a few seconds longer, all the while continuously observing the target, almost spot-welded to my cheek piece to hold my position steady. I stayed that way for several more long moments observing through the scope, then chambered another round and put my safety catch temporarily back to safe, then back to fire and I was back to observing the target for what seemed like a good couple of long, slow minutes. My mind was blank for a second, then all at once my legs started to come back to life and I felt pins and needles in both my feet and a numbness in my thighs and thought 'Oh shit' as I tried to move them. My normal drill for this was to wiggle my toes rapidly and try to arch my feet at the same time to get the feeling back into them. I needed to move my legs from underneath me.

These moments of discomfort passed soon enough, then I picked up my ejected case from the floor, wiped it on my trouser thigh and went to put it in my trouser pocket. At the very same time I heard a distinctive awesome sound: that of another .338 engaging. There is no other sound like it: perfection. One single pure sharp-sounding shot rang out above the noise of the now diminishing rifle fire that was going on around us: it was Ben, and with that one single shot, Ben had his first confirmed kill. We stayed up on the roof for another hour or so, observing and talking quietly but in great detail between ourselves about what we had just done.

Everyone was still on alert on the roof until the firing had completely died down. There was just the odd shot or two ringing out in the distance getting further and further away from our location and we were all now observing the area to our front and the compounds. The contact calmed right down and ended about four hours after it had begun. The clean-up on the roof soon started, checking for and repairing any damage to our defensive position where needed, with weapons being checked over and cleaned in situ on the rooftop. A couple of big green cans of rifle oil and flannelette were passed up in a large bag to clean the GPMGs and the 40mm GMG.

The gun crews were still sitting exactly where they had been fighting. Covered in sweat and dirt, hands and faces blackened with carbon from reloading the guns and conducting firing drills, the men talked among themselves with the odd sound of laughter and a smile. They were working hard to prepare the guns for stand to later that evening, or for firing in the next engagement should the insurgents soon mount another attack. A few sandbags and the empty ammunition containers were filled with empty cases and link from the various weapon systems that had been fired and were passed down off the roof and stacked up in piles, ready to be back-loaded on the next CLP back to Camp Bastion. Everything was back to normal in no time, with the usual stag rota now in force on duty on the rooftop, observing. The rest of the men were preparing their food for the evening, sorting out their administration and getting themselves ready for the nighttime routine after another mad day.

Ben and I were sitting down outside our living quarters in our improvised cooking area, making a brew and getting some food on the go. This was the usual: a boil-in-the-bag sausage and beans and lamb stew. We lived between two ISO (intermodal freight) containers which were filled to the brim with rations and stores. These were located at the gable end of a single-storey building in which the operations room was situated. The gap in-between the two containers was just over 3 metres in width, with hardboard over and above the gap. The boarded roof was sandbagged down to help prevent it being blown off by the wind and by the helicopters that landed on the HLS nearby. Our improvised roof also had a poncho covering it, for a little extra waterproofing. At each end we had built a sandbagged wall just above waist height to give us some protection from the sides when sleeping or just in case we might be caught out by indirect fire from the insurgents. There were two camp beds either side of the containers running head to toe in line with each other, each with the green dome mosquito net covering it. This is what Ben and I called home for several weeks.

Our food was nearly ready. The silver foil-like packets were bobbing about in the boiling water which was starting to boil over onto our Hexi and hissing away. On tonight's menu was Lamb Stew Navarin and rice pudding. I was just getting out the Worcester sauce and Ben had just started to make the brews when all of a sudden one of the guys came running over to us. He had just come down from the roof. He said excitedly: 'Monty, Ben – you got to come up to the roof now.' He then said: 'Monty, Monty, you know what you fired at? Well, get your kit on and come up and see quickly, something is happening in that area.' Immediately we both got up and grabbed our body armour and helmets, throwing them on, and grabbing our rifles and optics we made our way up onto the roof as quickly as possible.

I returned to my previous firing position on the rooftop from which I had fired that single shot earlier in the day, then hastily got out my spotting scope and binoculars and set them back up onto the target area. What we could see coming into our field of view in the distance and approaching slowly was an old red tractor driven by a young male. He was coming towards my previous target area, pulling a makeshift trailer with dark smoke coming out of the tractor's exhaust as it struggled slowly along the track.

There were two bearded elderly men on the back, holding on to the poorly-made wooden planks that were supposed to be the side supports of the trailer which was being thrown around over the potholes in the mud track as it moved sluggishly towards us. It then laboriously turned right, as if the tractor was struggling to make the turn, and drove past a small compound on the corner of the track. This was in full view to us. They proceeded further along the track and then stopped, with the tractor being blocked from view by the feature with the aperture on which I had been engaged.

The two elderly men then moved together from the rear of the trailer towards the rear of the tractor and disappeared from view for a couple of minutes. It was starting to get dark by now and the natural light was disappearing fast. Suddenly the first man came into view: he was hunched over slightly, facing forward as he moved slowly to the rear of the trailer. As he struggled to move, we could see he had hold of a corpse by the wrists. The lifeless body looked heavy, the head and dark hair hanging unnaturally towards the ground. It was dressed in what looked like a grey-coloured jacket and trousers with some sort of dark stained material hanging off the torso and dangling down.

At the other end, holding the legs by the ankles, was the second elderly man. Both were clearly struggling to carry the body for the very short walk back to the rear of the trailer. When the lead man reached the trailer they put the body down together. It was now out of view to us because of the wooden planks that made up the trailer's rear sides. They both then turned around and moved back the way they had come, so that both were now standing just behind the driver and holding on to the trailer's makeshift wooden sides. With that, the tractor started up again, smoke billowing out of the exhaust as the engine came back to life. It began to make its return journey, moving no faster than before, making the turn away from us at the same compound and travelling out of sight into the now darkness.

I kept my thoughts to myself but clearly remember I didn't feel anything about what I had just seen. We both packed up our kit in silence and came down off the roof. The regular nighttime routine had started; another long eventful day was at an end and moving into possibly another long night. For me this would include a short two-hour duty as watch keeper stag at some point: just me, with a radio and a computer monitor that illuminates where you are seated for company. There was always a continuous low background noise: the dull humming sound of the generator outside, powering the radios and battery chargers in the operations room that runs 24/7. At times there might be the odd crackle of interference or a voice coming out of the handset. This was when I would try to put events and actions into a logical order and then lock them away in the back of my mind. They would remain there as a colour replay: no sound, just pure images captured in my brain. At such times my tinnitus would be pissing me off because my surroundings were so quiet. But I am the lucky one: I get to see another day.

Ben's shot was confirmed earlier in the afternoon. There is no doubting the lethal effect of a single well–aimed shot at the target end from a .338 rifle, and now mine had been confirmed as well. We didn't say much to each other; we didn't need to. We had done our job that day, just as the other soldiers and specialists on the rooftop had done theirs, all working together. The observation post that I had engaged was collapsed by a couple of old men during the hours of daylight within forty–eight hours and completely disappeared as if it had never existed. For a while we did not receive any further incoming mortars or rockets, just the usual small–arms and RPG fire until the insurgents replaced that MFC with another who could adjust their mortars and rockets onto our location once again.

So the dangerous game, the whole process of finding that threat, started again with all players on both sides learning from their experiences and adapting, waiting for their moment to destroy or disrupt each other by sheer organized violence.

Chapter Four

Patrol Base Argyll

Part II

A fair few long days and nights had now gone by and the insurgents had continued to mount their attacks, both against our location and the foot and mobile patrols that went out from our base during daylight hours. This usually occurred in the early afternoon but they also began attempting to conduct attacks during the hours of darkness, normally somewhere around 2300hrs. Those night engagements were usually short and sharp. After a number of nights this seemed to be the pattern being set by the insurgents for their H-hour but the attacks were swiftly dealt with by the gun crews of the three tripod-mounted GPMGs in the sustained fire role. The Javelin crew would also launch a round or two into the darkness of the night, silencing the insurgent threats very swiftly and with pinpoint accuracy.

The various types of nighttime viewing devices included a thermal imager up on the rooftop with the Javelin (known as the CLU, or Command Launch Unit), giving the Royal Marines command of the darkness during which the sentries provided cover while the remainder slept. The regular exchange of tracer incoming from the insurgents and outgoing from the rooftop was a sight to see; I loved watching it and always have. The red tracer would arch up slightly as it left the gun barrels and then come down, zipping through the darkness in what looked like a continuous red line heading towards the insurgents' firing positions. Some tracer would strike the ground or make contact early with something hard that would cause the round to ricochet straight back up and into the night sky, still glowing red, till it burned out of sight. A cracking sound – zip, zip with every round – would pierce the silence of the night.

The sound of any activity around us and in everything we did seemed even louder when fighting at night. The darkness appeared to amplify every noise, making it seem even more hostile, aggressive and threatening, from commands being shouted out and passed along the rooftop to the sounds of the impacts, detonations of the 40mm grenades as they landed on the ground with an awesome dull thud, an echoing crump in the darkness off in the distance, to the sound of the Javelin missiles being launched and making a noise like one giant hand-held illumination rocket, a distinctive loud whooshing on firing as the rocket boosters kicked in. The latter always happened if we were both asleep. I would usually wake up first, fucking sharpish, and slightly startled for a few seconds; switching from comfortably asleep in my softie sleeping bag to full alert mode in an instant. I would try to get my hearing defence on as quickly as I could, cursing to myself for the first few moments, while at the same time trying to wake Ben who could sleep through almost anything.

The passing of each day was now bringing a change in the weather; just enough for us to notice the small subtle changes in our environment. A slight drop in air temperature, now with more cloud cover crossing the blue sky like a flotilla of large slow-moving greyish-white ships as far as the eye could see, with the sun breaking in and out of cover as the clouds passed overhead. More important were not the light conditions but the wind: it had started to pick up strength, varying from a moderate to fresh wind in the early afternoon which increased further as the afternoon moved into early evening.

We would watch the odd dust swirls that suddenly appeared from nowhere in the late afternoon, whipped up and speeding along the open ground like mini tornados and then disappearing as quickly as they came, dropping the sand, dirt, decaying foliage and twigs that they had picked up in their few seconds of life. There was still some evergreen foliage scattered about our area of observation, usually near a source of water. Some long grasses also remained and kept their natural green colour. Even as the natural light started to diminish during the day, they still thrived in this harsh environment.

In the areas around the base of the compound buildings, the bottom of the high compound walls, the edges of the dirt tracks and roads that crisscrossed our area and the small junctions of the drainage ditches there was still enough foliage growing to potentially conceal an armed insurgent or even a basic explosive device. As this could be done with very little ground sign or disturbance, these areas were still of interest to me and Ben: 'If there is green, it's a screen – observe it.'

Some such areas were often filled to a varying depth with water for the farmers' fields, as were any other natural depressions in the ground. Wherever water might collect in the ground there would be foliage: small clumps of bushes, some of which kept their leaves, and various tall trees. These were all good natural wind indicators that we could observe for sudden direction changes and to aid us in judging wind strength in the middle and far distance.

We had been out on a number of foot patrols by now. These usually ended in a fire-fight with the insurgents at some point but were still good from a sniper's point of view for gaining information on the ground. For example, the irrigation ditches that surrounded our location – how deep were they? Could they provide cover from view and fire to the insurgent? And what was their field of view? Were there any natural obstacles that the enemy could use against us or that might be a suitable location to place a well-concealed IED? How much of us or our patrol base could they observe and what could they see? The front gate, for example: an enemy would want to know when patrols leave and return to base; also the strength and composition of these patrols, whether on foot or mobile in vehicles, and the direction that they were taking. Could they observe our Sanger sentry positions, the routine for the sentries (including doubling them at night), and the weapon systems in these positions? All information about our pattern of life or routine within the patrol base was invaluable to the insurgents in planning any form of attack against us. This could even include our response time or reaction to an incident mounted against the base, or to a call sign on foot or mobile patrol in and around the area. Another very important factor was the time taken for the Medical Emergency Response Team to reach us from Camp Bastion when coming in to extract a casualty. Equally significant was the location of our EHLS: what a prize it would be for the enemy to take out a helicopter and crew on a CASEVAC mission!

No information on an enemy is useless as long as it is timely and accurate. From the compound buildings themselves and the firing positions inside and outside the compound to the actual size of the firing holes or apertures including doorways and windows and the thickness of the walls, all were of interest. The roofs of the compound – could they support a man or two's weight if they climbed up for elevation? It was often necessary to gain extra height for observation and fire in the summer months when the surrounding farmers' fields and land were covered in long grasses and tall crops.

Conversely any feature on the ground, natural or man-made, that could be used by the insurgent to observe or fire from or which might give him the upper hand was of interest to us. The same old questions that you ask yourself time and time again when you see something of interest on the ground suddenly pop into your head. Would I use it? And how would I use it? What are the advantages and disadvantages?

Another day started with the usual routine of brew and a bit of scoff, while at the same time conducting a bit of personal administration. A wipe-over of my .338 and a visual check of the rifle and optics, ensuring that nothing was damaged or missing, then using the cleaning rod so that my barrel was clean, even if the rifle had not been fired recently. A light coating of oil was applied where needed, followed by the checking and wiping-over of my rifle magazines and also that the ammunition within was free of dirt and grit. Finally my sight was cleaned with great care, paying particular attention to the lenses: a very fine soft brush was used to brush away particles of dirt or sand, and then with my fingertip wrapped in a lens cleaning cloth I would polish the lens, starting in the centre and working in a circular motion so that minute dirt or dust particles would be pushed from the centre to the outside and then away. The focusing ring, zoom control and elevation and deflection drums were also freed of dirt and grit and made easy to move. My final check was that the scope was still firmly secured to the rifle body and had not become loose. This routine had to be conducted daily by me, Ben and everyone else in the patrol base at some point. Weapon-cleaning is a soldier's pain: if you could calculate the number of man hours that a soldier spends on this during his career, what a total it would be in time and effort!

The insurgents had started to manoeuvre into their offensive positions and were preparing to attack the patrol base with a much better co-ordinated and larger force than usual, as it turned out later that day. It was now mid-morning on a clear, crisp day. The ICOM radio started to come to life in the interpreter's hand as he sat with us on the rooftop. Voices foreign to our ears could be heard coming from the small hand-held device, causing the interpreter to become excited. He sat up, moving slightly to left and right, peering over the low sandbag wall, his eyes wide open and scanning from left to right. Pointing in the direction of a large compound slightly right of our position, he frantically tried to tell us in broken English what was being said, translating the information as it came through.

The insurgent commanders were talking, letting each other know when they were in position and ready to strike, even giving us a little hint as to which weapon system they were going to use first. They too had their own form of code words, some of which were used for their more specialist weapon systems: the RPG had its own code word, as did the rockets and mortars. The message had been passed round the whole patrol base and everyone was either moving to their position or already in their stand-to position. This was a well-practised drill and happened without fuss so as not to give away our state of readiness. Most of the time the enemy would attack and even if we had Fast Air up or Apache gunships in the area they would conduct a very quick exchange of gun and RPG fire and then disappear as quickly as they came, back to their own positions of security and blending back into the local population until the next time.

The three GPMG SF guns were manned and ready to fire with the spare ammunition containers stacked up at the rear of the position ready for use; oil cans and flannelette were by each of the guns, along with the spare barrel bags and barrels. The Royal Marines were ready and waiting for the first fire control order to come from the gun controller who was on his knees with his binoculars, scanning the ground out to his front looking for a sighting or just a glimpse of the insurgents as a target. The Javelin crew, who were in the centre position on the rooftop as normal, were starting to pick up thermal signatures on the CLU as the insurgents came closer and they passed the locations of these to the guns. The 40mm grenade machine-gunner was standing behind his weapon, ready to fire. He traversed and scanned across the general area to his front with the launcher, observing the ground in detail as he did so. Two men were kneeling just to the rear of the GMG, ready to help with the reloading of ammunition and with target identification.

All this time the ICOM seemed like it was on permanent send, with the sounds of agitated voices coming out of the little radio and the interpreter trying to relay everything that was being said back to us in his broken English but in an alarmed state. At this point I was in a variation of the kneeling position and observing through my binoculars, looking in depth at previous firing positions. Ben was next to me doing the same with the spotting scope and using an old ammunition container with a half-filled sandbag on top as a seat.

I put my binoculars back into the claymore bag used to carry them and slung them over my shoulder, pushing them around onto my back out of the way. Things were starting to heat up. An attack of some sort was imminent and I put in my ear defenders, pushing them into my ears with a twist to ensure that they were correctly in place. I hated doing this: it seemed to amplify my tinnitus but better to do this than go deaf; my hearing has been damaged enough already. I adjusted my helmet and built up my position with my rifle, removing its bipod legs and putting them away in a pouch using a good old trusted trouser-twister to close them together and secure them away.

I got out my bean bag and put it on the sandbag in front of me, pressing it down firmly, making a small V-shaped depression in it with my hand so the rifle stock would fit snugly into it. This was my trusty old Italian camouflage material bean bag, which I had made some years previously and still used to do the job. Ben was in a sitting position next to me now, observing through the laser range-finder, giving me a running commentary on what he was observing and recording everything. We were double-checking our previous recorded data and areas of interest, paying particular attention as always to the wind direction and strength.

As usual within a few short seconds it all kicked off. Rounds from all directions came into our location, smacking into the sandbags and the wall below us and into the wall of the old school house. Insurgent hot lead was being fired towards us, rounds going over the top of our position and making a zipping sound as they travelled, snapping through the air above our heads. Dirt and small pieces of hessian were coming up from impacts with the sandbags. The silence of the day had been shattered in an instant. The fire-fight had been initiated by the enemy: they wanted a fight, and with the crack and thump of rounds being fired and impacting all around us, today's fight against us had started.

Within moments of the contact starting, two RPGs flew straight over us leaving small smoke trails in their wake and exploding in the air just behind us; fucking close, luckily no one was caught out. Blokes started to ping things – seeing things because of the occasional muzzle flash, a little smoke and some movement at the insurgent end as they manoeuvred in and out of their firing positions – just enough to get some initial fire back down towards the enemy. As this was going on, the gun controller located his first target through his binoculars and gave out his first fire control order of the day, the first of many. The guns started to spit out ball and tracer rounds at a rapid rate of fire, 7.62mm back towards the insurgent firing positions, ripping up the ground, the foliage and anything else they made contact with. They smacked into the compound walls and peppered the area, spitting up dirt and small clumps of dried mud into the air on impact.

The 40mm GMG opened up once more, spitting out its HE content as the gunner started firing at and engaging targets in and around a large compound located forward right from our position just over 500 metres away which had just opened up on us. Thud, thud, thud as the grenades were fired one after the other, a pause, and then again thud, thud, thud. The empty grenade cases and link were being rapidly ejected, spat out from the launcher onto the floor and starting to pile up underneath it. To all the sounds of sheer violence that were going on around us would soon be added that single shot fired from an unsuppressed .338.

The whole area to our front and the compound buildings within it were a sight to see; smoke and debris being thrown up from the ground by the impacts and detonations of various munitions finding their mark. The odd little random fire started to spring up in the foliage out to our front where the red tracer had impacted and still fizzled away ferociously, setting light to the grass around it before finally burning out. Back on the rooftop the familiar smell of cordite hung heavy around us as we tried to get some air into our lungs. Caused by the rapid rates of fire being used by the guns and by now one or two smoking barrels, the wind blew the stuff across the front of the rooftop, filling our nostrils once again. This was now becoming all too familiar to us.

I was observing the ground through my rifle scope, using my alternative kneeling position with my right ankle and foot acting as a support for my right arse cheek: with a slight movement from my right foot I could change my position slightly to raise or lower my point of aim and it was comfortable for short periods. My left foot was flat on the floor facing forward and my knee almost level with the right side of my face. I was now hunched slightly forward with my left arm fully outstretched, using my knee as a prop, and onto the sandbag as support.

Cupping my hand round the plastic stock on the front of my .338 I pulled it back into my shoulder and scanned very slowly across the front of a known insurgent firing point within a compound in the distance, a point that had been previously PID for regular firing and observation. Scanning along the front to the doorway, I looked into the dark backdrop of the entrance and moved along to the low windows, then to the corners and gaps in a low-lying wall which ran along the sides of the compound.

My magnification on the scope was on x14 and the field of view was as good as I was going to get for now due to the low cloud cover that was moving quickly across the sky and causing a slight deterioration in natural light conditions at this range (I could have turned the magnification down to let in more natural light but I was happy). I moved my head momentarily away from the scope, just long enough to blink a few times, refocus and get some moisture on the surface of the eye. I scanned back along the wall, observing each of its gaps in turn, then on to what appeared to be a hole larger than all the others.

This hole was a couple of feet up from the base of the wall and with a slight change in position I had good eyes onto it. All of a sudden two insurgents were coming into my field of view as I focused on the gap, moving from the right. Instantly my heart started racing and pounding, my body's sweat tap turned on almost immediately. Where the fuck did these two come from? All that went through my mind in that split second was 'Oh shit'. Then more importantly, what's the range? Check the range and confirm it, quickly. Can't fuck this one up, just because I was too idle to check my range card or use the laser range-finder. So with a quick glance down at the card I confirmed the figure in my head. It was more or less 770 metres to the wall and to what I was observing. They were close enough and I had no time to fuck about; they could both move off and just disappear out of sight at any moment.

Both men were in the standing position; the lead insurgent of the pair who was standing on the left of the gap was just slightly taller than the one on the right. He had what looked like a light machine gun or belt-feed LMG (an AK family variant) and was trying to set this up on the wall, holding the weapon at the butt end. The other guy was trying to help him by setting up the bipod legs of the weapon. I raised my head slightly from the rear of the scope to check my elevation drum; this needed to be corrected and some increase in the range added to its current setting which I always keep on 600 metres till an exact range to a target is given. I quickly made a slight adjustment so it was correct at 770 metres. Then a quick glance towards the right side of my optical scope at the deflection drum and I was happy with my wind correction. There was no time to waste.

Looking back through the scope and watching them both, I adjusted my focus control slightly for absolute clarity of the target sight picture, then using my right thumb I pushed the safety catch slowly from the rear fully forward, going from safe to fire. My heart pounded so strongly it felt like it might jump out of my mouth, and my mouth went really dry. As always at this point I had to fully concentrate and focus: try to block out everything that was going on around me. Just me and these two insurgents: we are the only people in the world right now; there is nothing else.

I then placed the centre of my crosshairs onto the first insurgent's head as he was facing forward towards me. He was still in the standing position, still holding the rear of his weapon with both hands and trying to keep it upright while what appeared to be his No 2 was trying to extend the legs of the bipod or release them. Once the LMG legs were sorted, the No 2 moved slightly away out of my sight picture, giving me a clear shot onto the main threat at the time. I superimposed the centre of the crosshairs just under the target's nose, on the fleshy part between the top lip and the base of the nose. This view and the remainder of his head filled my sight picture as I turned up the magnification of my scope to x18 and made minor adjustments to the focus. Then, with slow, controlled breathing, I maintained that sight picture while simultaneously applying pressure to the trigger and gently pulling it to the rear with the tip of my index finger.

BANG! In a split second the projectile was released and on its way to the target end. The vibration of this action was immediately felt on the side of my face and reverberated through my right ear defender as I rested the right side of my face on the rifle's cheek piece. All the while I was still observing, waiting for impact at the target end. These few moments of waiting and observing always seem to me to last for interminable minutes rather than seconds.

Then, smack! Impact at the target end to the front of the head where the round made its entry point, impacting and penetrating the soft fleshy tissue, entering cleanly and forcing the head violently to the rear with invisible force. The rear of the head looked like it had just exploded off; a red mist effect now filled my sight picture. A bloody red mush, a mixture of bone fragments, tissue, dark hair and cloth splattered out to the rear of the now opened-up skull in a flash. What was left of the head came back forward and slumped down towards the chest, now out of sight to me as the body and its weapon immediately dropped to the floor behind the cover of the wall.

Immediately I operated the bolt handle to eject the empty case as quickly as possible and chamber another round so I would be ready to fire again. I instinctively realigned onto my next target with a minor adjustment of my position, then placed the centre of the crosshairs onto the centre of observed mass of the second insurgent's body area. He was now facing me head on, still upright in the standing position with his weapon facing towards us: there was not a moment to lose. I just aimed and fired instinctively, waiting and watching as the seconds passed by, straining my right eye while observing and concentrating on the target. Sweat was pouring down my forehead, over my eyebrows and into my right eye, every drop of sweat giving a little stinging sensation.

The round impacted, making contact and entering the insurgent's lower jaw area just millimetres below the left side of his jaw. Instantly and violently this took a chunk of flesh and bone away from the side and back of his neck or base of the skull and went high left away from his body. A bright red mess of blood and tissue splattered outwards as if it had been torn off and his head snapped to the right as he fell to the ground. Again I quickly chambered another round and continued to observe the hole where the two insurgents had been for any signs of life. I thought: 'Fuck, my round was off slightly and dropped a little.'

I watched and waited but nothing or no one appeared in that firing hole. I managed to wipe my face while still observing the area. Even the sleeves of my shirt were wet with sweat: I was soaking all over and just wanted to take off my helmet to get some air to my head and pour some water over it. However, the fighting was still going on all around me and suddenly my little part of the world opened back up to me. Back to reality. Ben was on it, verbally giving out another target indication to the GMG gunner; a target that would need the attention of the 40mm.

Noise, smell and physical discomfort all came flooding back to me. I had to get my arse off my right leg which was now going numb and the familiar pins and needles were tingling away in my foot, as always after a short time in this firing position. I leaned back, stretched out my leg slightly and tried rotating it, moving it quickly up and down to get some life back into my foot. At the same time I reached out to pick up my two ejected cases and put them in my trouser pocket, and then I was straight back up and onto my knees.

Retrieving my binoculars from their claymore bag, I started to scan the ground to the right of our location; an area which was heavily grassed in places and had a prominent compound. This seemed to be the new direction of the insurgents' incoming firepower and they were on the move. Above us now were two Apache attack helicopters on station joining in the battle and hovering at a distance, observing and manoeuvring around in the sky like birds of prey. They would locate a threat and then destroy it with bursts from their main weapon systems ripping into one of the compounds used by the insurgents.

We continued to observe areas of interest and started to pan further right using our binoculars and spotting scope around the areas of some of the compounds and then concentrating further out on the flanks and into the far distance. We were trying to observe for the controllers the insurgent commanders as they gave their orders by radio to their fighters on the ground. These were the priority targets to find and take out. Over an hour and a half had now passed by since the start of the fire-fight. We had to look hard, sometimes very hard, just to find them: they are not always that easy to locate and take out; some are more experienced and better trained than others.

There was a small copse of trees sited just over 500 metres away and slightly right of our location that we were observing: some fire was starting to come our way from that general area and it was starting to increase. The trees were over 40 feet in height and swayed with the motion of the wind, their leaves still very lush and green at the top. At the base of the trees and at ground level the grass was long in places; a mixture of lush green and dying yellow which moved in unison with the trees above as the wind started to pick up again.

The light conditions were not improving either. Thick, puffy, greyish clouds started to fill the sky and in places obscured the sun from view for a good few minutes at a time. While scanning through our optics we caught some movement in the long grass: a figure in dark clothing wearing the familiar tan-coloured canvas chest rig that stood out from everything else and had initially caught my eye. He was kneeling next to one of the trees and leaning against it, using it as a support. There was an RPG warhead in one of his hands and the RPG launcher in the other; he was clearly preparing the rocket for firing.

Immediately I put my binoculars down between me and Ben and picked up my rifle, placing my trusty bean bag onto the solid sandbag and setting my rifle down on top of it, pushing it down as I did so for a little more cushioning and stability. I adjusted myself onto the new target: sight alignment and picture were good with a clear line of shot onto the mark. At the same time I was bringing Ben onto what I had seen; he picked up the binoculars and started to use them.

Both of us had now identified the target and there was no time to waste. I thought the range of 500 metres no problem; my elevation drum was set to 500 metres. As for the wind, a bit of sway but no great change in strength or direction. With not a moment to lose I brought the magnification down from x12 to x6, allowing a little more natural light into the scope, and with a minor adjustment to my final focus, I was on. The insurgent was still handling the warhead rocket, trying to get it into the top of the launcher. Steadying my firing position and controlling my breathing and sight picture for the final time, I couldn't waste another second: soon he would be able to fire the fucking thing.

I was aiming for the centre of observed mass of the target which was his chest area. I breathed out slowly and deliberately, bringing my crosshairs from above his head down and through the centre line of mass to the centre of the chest area, held that point of aim and then squeezed the trigger gently. My rifle fired. I held that position and sight picture for a few moments, then saw what looked like a small puff of clothing come away from the man's body as it was forced violently rearward, reeling backwards into the undergrowth, releasing the rocket warhead and launcher from his grip.

A small red puff of mist came off his chest and upwards: a mixture of flesh, blood and material from his clothing came away from the body as he fell backwards into the long grass and out of my sight. We watched where he had fallen for any signs of movement in the grass or any signs of life: there was nothing. I continued to observe this area for about ten minutes but still there was nothing, no movement. During those ten minutes things around us were starting to calm down, and judging from the ICOM chatter, the insurgents were apparently starting to withdraw.

We were just on the three-hour mark with this contact. My mind was clear at times but the image of the round impacting into one of the insurgent heads kept momentarily popping into my brain. I broke from my position, chambered another round, applied my safety catch and picked up the ejected case. My bipod went back onto my rifle and my sight covers back onto the ends of my optical scope: they were snapped off some time ago but had still done the job of protecting the optics.

At last the time came when we could both remove our helmets and undo our body armour. What a relief it was to once again be able to feel that cold air, a refreshing breeze on your head and body, to be able to scratch that itch, to rub your head from where your helmet has been resting securely on it and untangle your sweaty hair, to remove the sweat-soaked shirt with its white salt marks around the armpits back and front that stinks to high heaven, and to put on a dry, clean sweatshirt. Now that's an awesome feeling after a crazy day. It's the very small things, the little treats in life, especially when living on a patrol base, that you look forward to at the end of a shit day as a little morale-booster. For the two of us, it was usually food.

So far this had been our longest and largest contact with the insurgents, who are usually a lot harder to locate and observe. They use similar military skills and drills to ours: for example, cover from view and fire, clear arcs of fire, covered approach and exit routes, maybe even some form of rehearsals before an attack. The list goes on.

The sun started to go down on yet another long day, disappearing behind the cloud cover to be replaced by a very bright full moon, and the air temperature dropped enough to cause a slight chill as the evening began to set in. The day ended just as it had started with some rifle-cleaning, sorting out of personal equipment and the cleaning-up of our firing positions ready for the next engagement with the insurgents. This was followed by a cracking big mug of tea and some hot scoff: boil-in-the-bag sausage and baked beans with Tabasco sauce stirred in; the food of spartans. And another day closer to coming home.

Chapter Five

Patrol Base Argyll

Part III, Operation SOND CHARA

S everal days had now gone by and Ben, my No 2, had been replaced by Carl H. Ben returned very reluctantly to the reconnaissance platoon, where he was needed as a commander for a forthcoming operation that was to be mounted very soon and in which we would all be involved: Operation SOND CHARA (RED DAGGER). Carl H brought with him his L96 sniper rifle, taped up and sprayed up with great care and attention to detail; he himself was just as prepared and ready to go as his equipment.

My own rifle needed a little bit of attention by now; the scrim around the optical scope was looking rather the worse for wear. The frayed light-coloured hessian-type material that I had cut up into thin strips and wrapped around the scope was now becoming loose and stretched around the sight. It was secured by a thin light-coloured knotted string that I had cut up into pieces for the purpose. The string started life as a complete helmet net, one that I had removed from one of my Second World War British Tommy helmets before the tour. I kept the remainder as spare and carried it in my equipment.

I introduced Carl to three comrades from 35 Royal Engineer Regiment who had moved in with me and Ben a couple of weeks before; they had squeezed in between our two ISO containers and lived with us. They were constantly improvising and making things as Royal Engineers always do, so our living conditions were a bit more comfortable and weatherproofed. They ragged me for my attempt at sand-bagged blast walls that I had constructed together with Ben at both ends of the ISO container for a little extra protection from insurgent mortar and rocket attacks. There was L/Cpl Richie D, a very wise man for his years; Cpl Andy F, a true proud Scot, much of whose conversation I could never understand; and Sapper Eddie M, or Mally the Barmine kid as we used to call him.

These guys later in the operation were at the forefront of the EMOE teams, which involved going up and placing explosive charges against a wall or door and blowing it to make an entry point. This allowed the fully-equipped Royal Marines of 42 Commando access to the buildings or compounds to fight and clear insurgents from the small towns and villages within the Nad-e Ali area.

Carl and I filled the next few days preparing ourselves and our equipment for the forthcoming operation, which was now just days away. During the hours of daylight I would familiarize Carl with the daily routine of the patrol base, any areas of interest out on the ground from previous incidents, and also known previously used locations of recent IED strikes. Carl studied our mapping and familiarized himself with the spot codes for this particular area; also the very few available air photographs of Shin Kalay, Khushal Kalay and Zarghun Kalay areas that were up in the operations room. These were our possible destinations.

More important was a ground orientation, again requiring many hours spent on the rooftop conducting observation of the general area around our location. Then foot patrols, on occasion being able to show Carl the ground in detail from previous contacts, engagements against the insurgents, the compound locations of these firing points, known insurgent approach routes and favoured ambush locations for foot or mobile patrols within our area. Such information could be of great use to him for future reference while operating in this area.

We both started to pay particular attention to the changing weather conditions as the days passed; to the variations in light and wind throughout each day as it rolled into a long dark evening. The warmth, strong natural light and long days of the summer season were disappearing fast and conditions had begun to change for the worse in all respects. Winter was starting to creep in slowly, firstly denoting an increase in wind speed and possibly a change in direction. This meant that conditions became unpredictable until we studied the wind in detail. Also by this time the changing seasons brought the odd sandstorm or dust cloud that would suddenly appear from nowhere, only to vanish just as quickly as it came.

Rain or snow would eventually arrive, possibly with a low cloud base diminishing the natural light conditions and possibly affecting observation at greater ranges; also causing a drop in the air and ground temperature.

So it was worth re-checking and updating all previous data collected on the range card. For now, our information was as up-to-date as we could get it. It did not take long for the rain to arrive and when it did, the ground everywhere rapidly turned into a massive quagmire.

We always had to be thinking ahead, of anything and everything possible that might affect observation of an area of interest or a potential target and taking that final shot onto it. That is a sniper's priority at all times: neutralizing a threat to friendly forces when they are out on the ground, even if the friendlies themselves have not yet identified that threat. The sniper's actions must be quick, decisive and accurate in immediately neutralizing such threats, i.e. one shot, one kill.

Planning and preparation for the forthcoming Operation SOND CHARA was building up momentum as each day went by. Kit and equipment started to pour into our location on the CLPs daily and would arrive under the cover of darkness. Ammunition of all kinds flooded in on the backs of the large wagons and was unloaded, crates and boxes of the stuff: a mixture of 5.56mm, 7.62mm, 40mm HE rounds for the GMGs and AT4s as well as the LASM which is a similar weapon. Mortar rounds for the 81mm, smoke rounds, illumination rounds and HE rounds were piled up ready for use by the three dug-in 81mm mortar firing pits which were starting to fill up with rainwater despite the mortar crews' best efforts to keep them free of it. Four 105mm light field guns manned by the Royal Artillery had also reached us and the gunners were busy servicing and preparing the guns for firing and sorting out the ammunition ahead of the upcoming operation.

HESCO, to help provide protection from enemy fire, was going up everywhere and had been for a few days prior to the guns arriving, thanks to a section of Royal Engineers from 35 RE Regiment. Construction went on twenty-four hours a day in all weathers in the now very sodden earth. Small teams of Royal Engineers worked in twelve-hour shifts continuously round the clock and in the constant rain and mud under the watchful eye of Staff Mackie, even on occasions under small-arms fire, usually when the digger literally went outside the patrol base's walls to get more earth to fill up the HESCO baskets. The insurgents would have a go at the digger and the construction crews that just carried on and on. The gunners with their 105mm light field guns were also kept busy conducting firing drills day and night, constantly improving their positions around the gun line and the small purpose-built ammunition bunkers. There was constant activity everywhere within the patrol base and on the outside perimeter. Rations and bottled water arrived by what looked like the ton; medical kit and equipment was unloaded and stored in pre-designated ISO containers around the base.

Anyone and everyone who was available at the time helped to unload the trucks in work parties in lines of men from the vehicles, passing the equipment along the line as they stood in the mud and rain side by side, passing it down until it found its way to its final location as directed by the CQMS. This was so the vehicles could get away as quickly as possible under the cover of darkness with their Viking escorts providing protection. Some of the convoys had to fight their way in and out of our location at times. As the days passed and gathered momentum the convoys were regularly targeted by insurgents, so in addition to the Viking vehicles the convoys now had Scimitar escort as well from 1 PWRR Reconnaissance Platoon providing additional firepower.

Time was now running out and all the pieces of the game plan were nearly set for the operation, ready to launch forward into the insurgent strongholds commencing with a dismounted night attack. It was just days away and still tons and tons of fighting equipment, many men and a few women, all from different fighting units with different skills and trades, were massed into one place and with one unifying purpose: to take the fight to the Taliban on their ground and in their strongholds; to deny, disrupt, destroy, and take back ground and villages from under their control.

It was roughly midnight when yet another long convoy rolled into our location with its Viking and Scimitar escorts, churning up the mud around the compound as they made the tight turn around to face back out towards the direction of the front gate and lined up in single file ready for the long return journey to Camp Bastion. The big-drop vehicles and their cargo came in like great slow mammoth beasts, their big wheels straining through the thick deep mud, heavily laden with kit and equipment, and parked wherever the driver could squeeze the vehicle in and wait to be unloaded. It was good to catch up with Jimmy P and Andy L who were the commander and gunner of one of the Scimitar escorts and who I had not seen for about a month. They brought up our mail, which we had not received for a while, and some extra goodies. The mood was happy and relaxed with some banter and laughter on how the rest of the platoon was getting on and they told us about the contacts they had been involved in as we sat on the metal front decks of the Scimitar, still warm from the engine running.

Then Jimmy had some bad news to break to Carl: the news that he had to return with them to Camp Bastion and get ready to re-deploy with the platoon on the up-and-coming operation. Jimmy had to take Carl back as he was needed as a gunner in one of the Scimitars. The cycle of R&R had kicked in, leaving a gap or two in the platoon's ORBAT strength, and Carl had to return with them tonight. He was gutted, absolutely gutted, and didn't even say a word. I could see it in his face and body language, and there was nothing I could do. I could only mention that there was still plenty of time left, still three or four months left of the tour and plenty of time to get some sniping experience under his belt. The Taliban insurgents would still be here and were not going away just yet.

Carl went back to our living space between the ISO containers, to his bergan [rucksack] and equipment, and packed his kit in silence. Once again everyone was up and helping to unload the vehicles of their stores and ammunition, and the CQMS was running out of places to stack and store everything in a dry location. A few hours passed and the job was done. The vehicles were all lined up in their order of march; some drivers were conducting or completing the final checks around their vehicles ensuring they were ready and able to move off.

Often a vehicle or two would break down after so long on the unforgiving ground; maybe a problem with the engine overheating or the wheels or shocks of the vehicles. The tracks on the Vikings and Scimitars could be thrown or damaged very easily when constantly driven over the arduous rocky terrain by day and night and would need to be repaired as soon as possible by the crews. In a hostile environment, such as Nad-e Ali at the time, if you stayed static long enough the insurgents would be sure to attack if they thought they could take you on. The unfortunate crew and their vehicle had to get out of that situation and were usually hooked up to another vehicle to be towed or dragged back to Camp Bastion or the nearest PB.

Eventually the convoy rolled back out into the darkness of the night, the sounds of the various vehicle engines filling the previously silent night, as did the strong smell of exhaust fumes when they first moved off. As the convoy moved away and gained some distance from our location, the long dark vehicular snake disappeared from sight and sound, swallowed up into the darkness, and our regular nighttime routine kicked back in again. All that could be heard in the gloom was the low continuous hum of the generators in the background, powering the operations room radios and providing light for the duty watch keeper's small room. He maintained a listening watch and communications with the sentries and other locations on duty while the remainder of the base got their heads down.

Everything was now set for the forthcoming operation: kit, equipment and the troops to task, manpower to do the job; all was in place. The forces involved had been gathered, received orders and conducted rehearsals. There was heavy armour support from the Danish with their awesome Leopard 2 main battle tank with a 120mm smoothbore gun as its main armament, plus support from the Estonian forces, the ASF (Afghan Security Forces) and other members of ISAF (International Security Assistance Force).

It was a very damp and cold night with no cloud cover at all; the night sky was completely filled with small bright stars, what looked like hundreds or even thousands of them. There was a full moon that looked like it had just been literally hung there, high up in the night sky, beaming down its natural ambient light towards the Earth.

It was time to move out and get to our overwatch and start positions. I had got a lift at the start of Operation SOND CHARA in the back of one of the FSG (Fire Support Group) WMIKs (a Land Rover which can be fitted with either a 40mm GMG or a .50 calibre HMG and also has GPMGs fitted on the front) from 42 Command. Mel was the FSG commander of this very experienced group of Royal Marines. I sat, or tried to, in a kind of small metal bin on the back with my legs inside and my arse hanging over the edge, hanging on to the metal frame of the vehicle every time we went mobile. There was nowhere else for me to go without getting in the way of the vehicle gunner. Anyway, it would not be for long.

My .338 rifle was in-between my legs and A2 rifle on my back, with my day sack secured to the side of the vehicle. I had stripped down my kit as much as possible, right back to basics. Ammunition: no compromise; if I can move with it then I must carry it, as much as I can. This included two HE grenades, two smoke grenades and all of my six issued .338 magazines, fully charged. As my reserve I had an H83 container full of a mixture of loose and boxed 8.59mm ball ammunition, this carried in my day sack readily to hand, with my twelve magazines of 5.56mm in my kit.

My optics included my German 8x30 Hensoldt Wetzlar binoculars which were an excellent lightweight set, a small spotting scope, my issued laser range-finder and my trusty pointer staff, plus the basic medical kit that we are all issued and must carry. As for food and water, I carried enough for forty-eight hours. My personal kit comprised a softie jacket and trousers, Gortex jacket and trousers, a Bivvi bag and snug-pack sleeping bag; also my half body-length blow-up Thermarest mat, three pairs of socks, my trusty old faithful Falklands hat and green face veil.

We moved through the darkness of night with caution; often stopping, listening and watching using our nighttime viewing devices to maximum effect in total silence. At anything suspect or of interest we dismounted from the vehicles and investigated. I remember I was only wearing my under-armour shirt because we might have to fight our way there and into the village, and the last thing you need is for the body to start overheating while in a long and large engagement. It is always easier to warm up than to cool down. The shirt was soaked with sweat and every time we stopped for a short period I would feel the slightest breeze sending a chill through me; every now and then I would use the sleeve of my shirt to wipe away the sweat from under my helmet.

I could not wait for the sun to rise, which for me is a psychological thing; most soldiers do not enjoy the darkness. As soon as the sun starts to rise you feel just a little warmth, coming from the cold darkness of the night into the daylight, a bit of morale, and of course some real physical warmth projected down from the sun onto your skin.

Those hours of darkness seemed to pass by really slowly. We were moving, stopping, going firm again and again in short bounds for a while, listening and observing into the darkness then moving off slowly once more. In the distance and over the radio call signs were already in contact: the night-fighting had begun. From that point on, often heavy exchanges of gunfire could be heard echoing through the darkness.

Occasionally off in the distance the night sky would light up with bursts of tracer streaming and arching across the darkness, then coming back down to the ground. Friendly force illumination rounds from the 81mm mortars going up made a kind of whistling sound as they travelled high up into the night sky and then, pop: instant brilliant white light as they lit up the darkness and the ground below, creating shadows as they were carried along by the wind and then died out. The area all around us was then plunged back into complete darkness and silence until the next fire mission was needed by a call sign or there was another exchange of fire with the insurgents.

We had been in position for about three hours waiting for H-hour under the cover of darkness and had cleared and occupied a derelict building. We had managed to get to our location unchallenged and were ahead of time, so we waited. Other call signs were at this point in time still moving and making their way to their respective start locations or LOD, having already been in contact en route during the night. They were having to fight their way through to get to their start lines, even before the main advance into Shin Kalay could begin.

I was lucky, having two choices from my rooftop position: I could stand up to observe and fire or I could adopt the kneeling position instead. I also had room to allow for minimal movement either side of my chosen firing apertures to prepare my equipment, and a place to urinate away from my firing positions roughly 2 metres to the rear right, there being no need for a piss bottle on this operation. This location offered me good cover from view and fire, certainly on my left side.

On one side to my left were the remnants of a rubble-cornered gable end wall of the building plastered with shrapnel and bullet strikes; brickwork rubble and splintered pieces of wood covered the uneven floor, which I cleared easily enough. The route into and out of this position was pretty straightforward with no real hazards in the darkness getting up onto the roof. The stairs were just about usable and an entry point into the building had already been created by previous fighting and shelling. Also on the roof, offset from and behind me in cover from view, was the OC's grouping with a mixture of FACs, MFCs, FOOs and the remainder of his command group. It offered them, as well as me, good fields of observation so they could control and direct their respective fire missions when and where needed and called for.

I could also see across to my right a possible alternative firing position that might open up even more of my field of observation and fire into the town, should I need to relocate as things progressed. This was a small depression in the concrete floor of the roof, with a small pile of masonry rubble with thick wire support prongs twisting sharply out from it in all directions. This must once have been part of a large pillar support for the building. I might just be able to make use of that position later in the day as it would increase my arcs of view and fire further to the north-west. It would need a little working on to provide me with some protection from small-arms fire but I could build it up with some of the rubble lying around on the roof and also down below on the ground floor. I was starting to get a good feel for my new surroundings and our environment.

As the dawn of a new day approached, I got my binoculars and spotting scope ready to start scanning the area to my front and try to pick out possible insurgent firing points and locations and to use the laser range-finder to accurately get the ranges to these places and record them for future reference. The sun rose slowly into the early-morning sky and the dawn of yet another day in Afghanistan was soon upon us. But this day would be different. It was the day we all had been preparing and waiting for: the start of Operation SOND CHARA. The sun gradually spread its light and warmth over everything below as it climbed high into the morning sky, coming up over what would be our next battleground. I could see friendly force troops in their start positions and vehicles waiting to launch forward. Hopefully from this position I would have good, clear arcs of observation and fire into Shin Kalay.

My position was again atop a small two-storey building, or what was left of one. Going for elevation and a location hopefully not too obvious for a single sniper mixed in with other friendly forces concentrated in a small area, sometimes you have little choice. It did at least offer some protection from fire and view. There were parts of the building that had been turned into areas of just standing rubble. It was used by the insurgents regularly as a firing position onto our patrol base and against our foot and mobile patrols, offering them direct line of sight and fire onto us.

However, this worked both ways and for us it offered the MSG gunners a cracking target to engage with the three sustained fire role GPMGs, the odd Javelin round or two and the 40mm grenade-launcher. But time and time again the enemy would use this position and even left their empty cases and LMG link in location, together with small piles of bloodstained torn or shredded clothing in places around the building and blood spatters on the floor and walls.

Today there were a few large puffy white clouds hanging motionless over the town against a near-perfect sky-blue background, with a long white vapour trail or two streaming straight overhead where Fast Air had done their overflight some time earlier in the morning. The local wildlife started to stir, the usual barking of dogs could be heard all around us and every now and then startled birds took to the sky from a small tree line forward right from my position. They flew around in what looked like one big loop, going up high into the sky en masse, swooping low to the ground and then back to rest from where they had taken flight.

To my front I could clearly see the uneven, potholed road that split the town in two running towards the mass of man-made structures that must have been the town centre. I used this as my axis, initially to break the ground down left and right, also using reference points on some prominent features that I had picked out on either side of the main track for quick target acquisition. On the left side of the road were compound buildings and walls, all apparently running into one another with alleyways and entrances into narrow, tight streets. A perfect setting for the insurgent to shoot and scoot or ambush and use whatever resources he had at the time to kill or maim us, maximizing his effect against us by making good use of the compound buildings and their height advantage to overwatch and engage friendly forces moving through the town.

There were so many windows, doorways and rooftops – anything that could be used as a firing position – to cover but I had a good clear view covering the left-hand side of the town. To the right of the road was an open area of ground about 500 metres in length by just over 620 metres in depth; what looked like well maintained farmland. There was a recently-ploughed and muddy-looking field with water-filled irrigation ditches running along its sides, and a small tree line in the top right corner which ran out to a further 100 metres or more away and ended with a long, high, well-constructed compound wall which ran from the tree line back across the fielded area towards the muddy road. Behind the wall was another mass of interlocking compounds, buildings similar to those on the other side of the road; a potentially busy arc to observe over once it all kicked off. There were other call signs observing over the same area, watching and waiting for the command to launch forward into the town.

I could hear the noise of the Viking and Scimitar vehicles in the background as they waited to go forward. Just visible was the turret of the lead Scimitar, scanning the ground to its front, left and right. The small turret slowly traversed the 30mm gun, moving and then coming to a halt, slightly depressing or elevating the gun barrel while the gunner conducted observation through the sight onto a fixed point of interest from within the tight confines of the turret.

They were down to the right from my position. I was still in the standing position and, like everyone else, starting to warm up slowly from the nighttime insertion but I clearly remember my feet were still freezing wet. My toes, indeed the whole of both of my feet, were numb. An uncomfortable cold numbness pervaded all my extremities from the night before, trying to cross the drainage ditches, farmland and tracks, shin-deep to knee-deep in mud in places on the final part of the route in. And as always happens in the darkness to many a soldier, definitely to me, a little unexpected fall or slip into the mud, water or animal shit; maybe even a slight uncomfortable twist or jarring of the knee or ankle from your little trip. That really pisses you off as you get up and try to walk it off, continuing the patrol into the darkness, cursing to yourself under your breath and quickly checking you haven't dropped anything from your kit.

I had my rifle down by my right side at this point, leaning against the wall supported by my oversized day sack so it would not fall as I remained standing, observing through my binoculars to the right-hand side of the town. There was no civilian movement at all from what I could see. My bean bag was already on the wall in front of me for use as a firing support, to provide a cushion between my rifle and the wall, and I had already removed my bipod from the rifle and put it away in my belt kit. I picked my rifle up, moved the bean bag a few inches to the right and set my rifle down on it, roughly in the centre of the stock, then with a quick check of my elevation and deflection drums I started to use the scope for observation. I turned up the magnification ring to x16 and with a final adjustment to the focus, my sight picture was crystal clear. The natural light conditions were very good compared to the last few days. I began observing an area around the tree line by the compound wall forward right. It was a comfortable observation and firing position without too much strain on my right hand, arm and shoulder.

I slowly started to scan the area, going from left to right of a small arc that I had given myself overlapping the ground, paying particular attention to the area where the wall seemed to meet the overgrown foliage of the tree line. There was plenty of trapped shadow and screen: I thought you could probably hide a section plus of men in there with ease, if they used the irrigation ditch that went into the tree line. A gift of a firing position, I thought. Insurgent movement had been picked up by other assets and suddenly our radios started to come to life with chatter, with important information from other call signs on the ground and from the air. At the same time the ICOM radio used by the interpreters burst into life too. The insurgents were manoeuvring into their final firing positions and getting ready to attack.

Abruptly the silence was broken by small-arms fire coming towards the vehicles: a couple of single shots, then erratic long bursts. I looked down and could see the impact of the rounds dropping slightly short of the lead vehicle, spitting up mud and water into the air where they struck the ground and then seconds later the next few bursts found their mark, impacting on the side and turret area of the lead Scimitar. The sound of the impacts filled the air: dink, dink, dink as the rounds hit home and some ricochets bounced off the armour, whizzing off into the sky in all directions. Troops urgently taking cover behind the vehicles were looking and trying to locate the insurgent firing points now engaging them.

I looked in the direction of fire and scanned the area frantically, looking for any smoke or muzzle flash coming from the insurgent weapon systems as they were firing or just some movement in that area, anything that might give away their positions. Within moments a dull heavy sound – thud, thud, thud – filled the air with just a second or so between each round as the 30mm RARDEN cannon kicked in. With that came the clanking sound of empty cases being ejected from the front of the turret as they were fired, bouncing off the front metal decks of the vehicle, metal clanging on metal, and falling to the muddy ground by the vehicle tracks. Thud, thud, thud again as the Scimitar fired another three-round burst, then another and another in quick succession. As the firing continued, the 30mm cases started to pile up on the front decks around the closed hatch of the driver's compartment and collect on the ground.

The second Scimitar's engine growled into life with a large black cloud being coughed out from its exhaust as the driver put his foot down. The vehicle screeched and jolted violently, manoeuvring slightly forward right of the firing Scimitar and started firing its 7.62mm chain gun with long bursts, first ripping up the ground to the vehicle's front onto a forward edge of one of the drainage ditches in the field and spitting up mud and grass into the air along its trajectory as the rounds fell slightly short, then found their mark. The 7.62mm ball and tracer flew in what looked like a continuous line of red towards the tree line. Then one of the vehicle-mounted GMG 40mm grenade-launchers started firing, fire control orders were being shouted out, and men were moving around everywhere; moving to get eyes onto the insurgents and fire their rifles towards the tree line.

The FACs and FOOs were all trying to locate the insurgent firing positions, manoeuvring themselves around the roof area with caution as the incoming fire was now impacting over and around us, their binoculars in one hand and a rifle or radio handset in the other. They peered over the rubble to identify the positions and call in a fire mission to neutralize them swiftly. The right side of the roof was organized chaos with people firing everywhere, and a few more Royal Marines came up to join us for better eyes on the ground and also started firing their weapons. Empty cases were now littering the rooftop, with link from the GPMGs starting to form neat little piles underneath as the gunners fired burst after burst in the prone position. The individual gunners would only clear the link away when a fresh belt of ammunition was being loaded or a stoppage on the gun being cleared, returning to firing again as quickly as possible. All this tumult of noise and violence happened concurrently, picking up momentum and intensifying as the morning rolled on.

I was now concentrating on what looked like another couple of possible random observation or firing holes in a section of wall about 60 metres down from the start of the tree line as it met the corner of a higher compound wall. I scanned slowly up and down along the length of this wall with my rifle scope, and then along the top. A sudden movement caught my eye: I froze my position and held that sight picture for several moments, then it appeared again. I caught a glimpse of a man's head and Talib headdress in the scope as he peered over the wall with great caution and was only visible to me for a few seconds before ducking back down out of sight. I thought: 'I've got you, I've fucking seen you.' Here we go again, I said to myself; this guy is up to something behind that wall.

Several more long-drawn-out minutes passed by as I waited to see if he would reappear. Then two hands appeared on top of the wall, roughly shoulder-width apart, and the guy started slowly trying to pull himself up onto the wall. All the while he tried to maintain a low profile, keeping close to the wall and almost hugging it as he struggled upwards and then lay flat on his front. Once he managed to get the rest of his body up and onto the wall, he struggled to balance in this awkward position. He now filled my sight picture, and what a sight picture that was. All was crystal clear to me within those few moments of observation: who and why and what he was attempting to do; positively identified as a 100 per cent pure bad guy.

The insurgent was now lying in full view to me side-on: his head, the trunk of his body and his legs were all visible. Time, as always, was of the essence in this type of situation: act and act fucking quickly as such opportunities are rare. I could clearly identify that he was armed with an AK variant slung over onto his back; the material of his rifle sling stood out to me, light in colour and nearly matching the crude assault vest that he was wearing. The latter did not seem too secure around his chest as he pulled himself up and onto the wall and it was causing him problems, getting in the way of his efforts.

He was dressed in a pair of greyish, baggy, long trousers and a slightly darker long-sleeved undershirt. Over this was a brownish, thickly-padded, sleeveless jacket that was not fastened, with the assault vest underneath. His facial features that I could discern were well-weathered and mature but mostly hidden by thick, dark facial hair that met his hairline at the sides of his headdress.

I moved my safety catch from safe to fire, slowly pushing it fully forward with my right thumb, all the time trying to maintain my sight picture to focus on what I was observing. I was pretty sure from my previous calculations that his position on that wall was just over 630 metres away from me and I had no reason to break my position and check my elevation drum or my deflection drum. I was on, and at about 630 metres it was a gift of a shot. I had used my laser range-finder on the holes in the wall earlier that morning and this had put my mind at ease regarding the range.

As always at such intense moments I started to break out in a cold uncomfortable sweat as I tried to control my breathing, keeping it calm, with a long, slow, deep initial breath. Then I concentrated on slow, controlled, deep breathing, taking in as much air as possible each time, holding it for a second or two and then controlling my gradual outward release of carbon dioxide.

My eyes always needed a few blinks to clear them and get some moisture back onto the surface of the eyeball and re-focus on what I was seeing. The eyes always feel so dry; staring and observing through optics for prolonged periods of time in the heat and straining the eyeball. I would also tighten my grip on the rifle under the control of my right hand, trying to pull it back just a fraction more into the body armour that padded my right shoulder. My left arm was almost fully extended out forward, my forearm resting on part of the wall and the edge of the bean bag, my left hand cupping the front end of the stock, just lightly pulling back into the shoulder as well. My left leg and foot were positioned facing forward to the base of the wall using it for support; my right leg and foot were in direct line with the body, now my main support in this standing position. My heart began to pound strongly once again, every beat seeming to ripple out through the rest of my body.

The tip of my right index finger just touching the trigger relaxed slightly, while the rest of my right hand had a vice-like grip on the rifle. Everything for me was now set. I watched as the insurgent fell to the ground, landed on both feet and stayed in a crouching position. At this point I realized he must be in a shallow ditch that ran along the base of the whole wall towards and leading into the tree line off to the right. The ditch also ran back through the other way, leading down towards the road. The high mud wall being the direct backdrop to this ditch, I aligned the centre of my sight pattern onto the insurgent.

I observed him for several moments more and made a very minor adjustment on my fine focus, bringing him into absolute clarity within my scope and eliminating a little drifting mirage at the target end. My sight picture was now so clear and sharp. He turned his upper body slightly to the left, bringing the AK round to the front of his chest. Holding it firmly with both hands, he was slightly hunched over, then leaned forward and adopted the prone position, using the ditch to his front for support and cover. To me he was now clearly about to fire his weapon and constituted a threat. He then turned his head up and slightly to the right as if looking back up to the side of the wall. As he brought his head back down and in line with his weapon, I aligned my centre crosshairs onto the centre of observed mass of his head.

I could now clearly see and observe through my rifle scope the insurgent's weathered features, his mass of dark facial hair and the head hair protruding from beneath his headdress. My entire focus was on placing my crosshairs and holding that point of aim on the centre of observed visible mass. I held that sight picture just a little longer, speculating to myself who would be the first to fire out of the two of us. Very slowly I increased the pressure to my trigger finger. Then suddenly: BANG, that familiar sound and action of the rifle as it sent the round on its way to the target and that oh-so-familiar feeling as I sensed the power from that action in my right shoulder. The vibration could be felt through my right ear defender, having contact with the cheek piece on the rifle butt as my right cheek rested on it.

I still held the picture and seconds passed by, feeling like minutes. My mouth was so dry as I watched and waited; then in an instant: splat! The round had found its mark and made contact at the target end. I could see the violent jerk of the insurgent's head through my rifle scope as the 8.59mm round impacted. The lower wall behind his head where he lay was immediately covered in a pattern of blood spatters: a red mist of blood, tissue and what must have been bone fragments from the side of his skull. The spatter pattern seemed to me slightly higher than his previous head and body position, going up the back of the wall. Fleshy tissue and a clump of blood-matted hair dangled freely from the top part of his head. The head instantly fell forward, dropping towards the insurgent's rifle. I watched and re-focused onto this area as the lifeless heap lay there atop the irrigation ditch. Motionless, the top half of the body's trunk was still visible to me.

I immediately chambered another round, re-aligned my sight picture back onto the fallen insurgent and scanned the ground to left and right, then back onto him and scanned along the top of the wall. There was no movement, nothing. I wondered where his fellow fighters might be. Maybe on the other side of the wall, waiting, and would have heard the impact of the round as it penetrated their comrade and went into the wall, all in a split second. I wiped my face and neck with my sweat rag, picked up my ejected case and put it away quickly in my day sack along with my firing log book and ASATS cards.

Taking up my binoculars I had a good look at the rest of the area to my front, scanning back over the rooftops, compound walls, doorways, windows and prominent ground features because you never know; all that's needed is just a glimpse of something to make you curious enough to wait and watch. I was drawn back to what was happening to the small tree line on the right of the compound wall. It was an awesome sight, getting some heavy punishment from the Scimitars and Vikings now firing their respective weapon systems. There was the dull crump of 30mm HE from a three-round burst followed by another and another, impacting at the target end and throwing up dirt and foliage at least a couple of feet into the air. The chain gun of the second Scimitar and the Viking vehicles' GPMG 7.62mm ball and tracer were ripping up the earth around the forward edge of the ditch, red tracer streaming out from the guns towards the insurgent firing positions.

Re-attaching the bipod legs back onto my rifle, I clipped my rifle sling back onto the adjustable hand-stop (a forward sling swivel on the underside of the plastic rifle stock), and clipped it back onto the rear of the butt. Bringing the rifle back up, I started scanning over to the left side of the road that ran from behind me, roughly from the 6 o'clock position straight up through to 12 o'clock and out of sight to me. Then the friendly forces troops started to move forward into the town; men and vehicles surged slowly forward in single file. I could see two military dog-handlers with their animals barking and pulling forward on their leads, mixed in-between the advancing troops. I could also make out some Royal Engineers preparing to go with all the EMOE kit that was attached to their large day sacks and carried on their backs; they readied themselves to advance forward along with everyone else.

Time moved on and it was now nearing midday. Every now and then sporadic gunfire ahead and dull explosions off in the distance could be heard. On either side of the road which acted as the axis, the soldiers and marines would react and sure enough the threat would be silenced; they would then move on deeper into the town. Two large explosions erupted out into the middle distance roughly 600 metres away from me, just over on the left-hand side of the road within the now heavily-compounded area of the town buildings. Two light-brown plumes of dirt and debris came up from ground level, rising up 100 metres or so and looking just like mini atomic bomb explosions; a mushroom-cloud effect from within two compound areas climbing up into the early afternoon sky and just hanging there motionless with hardly any wind to disperse them. This was quickly followed by the varied sounds of smaller explosions, a grenade or two, a few bursts of automatic fire followed by several rounds of single fire in rapid succession, then silence. This was the sound of the coalition forces gaining entry into the compounds and systematically clearing them of any insurgents.

As they moved deeper into the town it was time for me to start thinking about leaving my present position for another secondary location. I needed to cover more of the town and get better eyes onto the left-hand side of it as the insurgents were now manoeuvring around in that area, some having been spotted on the outskirts by other assets. But my options were limited due to this not being a conventional setting and my having very little freedom of movement in this environment.

We all have control measures for a reason. As snipers we could be on the outskirts of a town or urban complex or dwelling in formation, ready to mount a close precision attack. We could be on one of the flanks of a town or village depending on the ground around this location, and must not get in the way of the assault bearing or direction of attack for the assaulting troops. We must then be able to locate and identify priority targets, engage such targets and also trigger information to other assets from outside the town, directing forces into the town to neutralize that threat. We must have some degree of freedom of movement, able to move out and along the flanks either slightly ahead of the advance or level, again engaging targets and reporting on what we see, with a FOO, an FAC or an FST attachment able to engage bigger targets and destroy them. This would range from light soft-skinned vehicles and light armour to crew-served weapons that may have some form of protection around or over them and any identified communications equipment; the list can go on and on. Being able to look in depth and provide timely and accurate information to the commanders on the ground on what lies ahead of them, or what we can observe that the troops have not yet identified on the ground such as obstacles and threats so that they may counter-plan and act, is essential.

All the time we would be indexing and prioritizing targets, updating information, engaging and observing threats and reporting what is seen back to the sniper platoon commander. He is in command of the screen on the ground, makes the decisions and passes the vital information up the chain of command. As a quick example, regarding obstacles observed or the location of enemy MGs, LMGs, bunker-type reinforced positions and possible command locations: are there any crew-manned weapon systems in static locations or are they mobile? In other words, almost any information regarding anything that may be useful about the enemy and may cause a drama to the assaulting troops is passed on in a timely manner. To do all this you must try to maintain good radio communications. An FOO, MFC party or even an FAC might be attached to the screen, in which case the old saying 'no comms, no bombs' becomes extremely relevant.

I made sure that the position I had occupied for the last few hours was clear and double-checked that everything I had used was secure and packed away where it should be. The corner where I had had to urinate was still damp, so I just kicked some dirt over it to cover it up. I came out of the observation and firing position that had kept me in the relative shade (trapped shadow as we would call it) for most of that morning and could now stretch my legs properly, walking back into the bright sunlight and out onto the open rooftop. I soon felt the warmth of the sun on my face and hands as I began to warm up all over, even my feet starting to feel normal again. I just knelt down behind some rubble and observed back out to my front for a few minutes, watching what was going on in the town. It seemed like a scene from a war film; however, this was real, very real.

I went over to investigate a small pile of rubble that I had been considering using as a secondary firing position. It used to be one of the main support pillars of the building and was now just a heap of rubble with a couple of thick, twisted wire support rods poking out. The arcs of fire and observation were very good from this location but it was obvious, very obvious, being off to the far side of the roof in an open area away from everyone else. However, it was now no longer about concealment and surprise as our forces along with the ANA were in and out of contact with the insurgents in the town, continuing their advance compound by compound and street by street.

I would have to do a little bit of building-up of the position if I wanted to make use of it. The rooftop was still occupied by our forces and some of the ANA were also up there, watching and observing what was going on and talking to their men on the ground in the town by radio. I figured the position would not take too long to build: all I needed was some cover from fire from the front and a little flank protection and the job would be done, providing a good short-term position. There was still plenty of rooftop action going on with soldiers everywhere, some bringing up boxes and crates of ammunition, others moving around trying to observe for in-depth insurgent positions, some firing. One of the command groups and their attachments the FACs and FOOs, all of them wearing their heavily-laden day sacks on their backs, were gathered around a map on the ground over on the left side. All were on their knees, bent over, looking, listening and pointing to various locations on the map and then coming up with solutions to the ever-changing battle on the ground. The radio operators of the respective groups would then get this information out over the radio net to the forward and flanking call signs, followed by a fire control order or two hastily being shouted out above the general din. Men quickly reacted to those orders with a firing or two of LASM being hastily launched and the sudden sound of small-arms fire all around us. A mixture of 7.62mm and 5.56mm would zip through the air, exiting the rooftop at a rapid rate.

I watched what was going on around me. Out to my front could be seen odd plumes of dirt and debris being thrown up high into the early-afternoon sky by the powerful explosions created by the Royal Engineer EMOE teams as they blasted entry points into the compounds for the Royal Marines to assault and clear these locations of insurgents. I could now determine how far they had actually advanced into the town.

Having collected some rubble from around the rooftop, I also rearranged some of the loose debris lying around the base of the pillar that I wanted to use and hastily built a sort of L-shaped low-lying wall that I could occupy in a slightly hunched sitting position, being able to observe and fire comfortably from here. It didn't take long to make and I got my arse down in there, adopting the sitting position with my day sack down by my left side. I could rest the bipod of my rifle on the rubble to my front and change my angle of fire from facing forward, using the road as my axis and dividing the ground in half again. I could then change my arcs of fire to either left or right, with good clear observation out to some distance. Visibility at the greater ranges was very good at this time of day, hardly impaired by mirage.

I had been concentrating my observation along the rooftops and the outer limits along the outskirts of the town, looking in depth but also into the middle and far distances at prominent features and areas of interest on the ground to the left and right of the main road. This was an area where insurgent commanders or observers could move about freely with no real immediate threat to them, able to direct their forces against us and control their side of the battle. They were usually highly mobile with expert knowledge of the ground they were operating in and made very good use of motorbikes or four-wheeled vehicles to traverse the general area, never remaining static for long. Usually they carried a radio or mobile phone to speak to their subordinate commanders on the ground in battle, plus some form of optics for observing us and our locations. A personal bodyguard or two, lightly armed for their protection, would also accompany them on the ground.

The battle was now moving forward, ground and objectives being taken by the coalition forces and what looked like a continuous stream of men and vehicles snaking its way slowly forward into the town. Medics with their oversized bergans jam-packed with medical equipment and supplies on their backs and the dog-handlers with their barking and very excited charges just wanting to be let loose were mixed in file among the fighting soldiers as they continuously surged forward. Apache gunships were flying above the town ready to strike at a moment's notice: they would circle above and then move away to stand-off positions; hovering, watching and waiting.

About three hours had now elapsed, during which time I had been observing and lasing a few potential positions and prominent reference points on the ground out into the far distance and plotting these onto my map. There was an ANA officer up on the rooftop with a set of binoculars and, like me, he was constantly scanning the ground to our front: every now and then he would point out something of interest and took up his position of observation next to me. He would get my attention either by voice or by tapping me on the shoulder, physically pointing in the general direction and then trying to explain to me in broken English what he had seen. I would then either use my rifle scope and turn up the magnification and focus on the object, or use the Leupold to confirm what he was looking at. After observing it and letting him have a look through the scope or Leupold we would quickly exchange our views and this went on for some time. Something then caught our eyes slightly over to our left in the middle distance: something not in that area from our previous observation and we both picked up on it.

We were observing a small part of a flat rooftop and from our location we had a full side-on view into this area of interest. What could be clearly seen through our optics was what appeared to be a straight line, several metres in length, and dark in colour compared to its whitish backdrop of the slightly elevated rooftop behind. It was pointing outwards, coming up from behind or below some sort of cover and at a slight angle. My first thought was that it must be some sort of antenna and after a few minutes of observing we both agreed on this. My rifle at this moment was set up in front of me, its bipod resting firmly on the rubble to my front. I was sitting on my arse with both cheeks on the floor and my legs crossed in front of me where the rifle body came back and down towards me, the butt itself resting on the lower part of my legs. I hunched slightly forward, observing through my binoculars with my elbows resting on the inside of my thighs, using them as support for my arms.

Putting the binoculars down, I built up my firing position onto what we were observing. My rifle butt was now firmly in my shoulder and once again I had to change my elevation drum, adding some clicks to increase the range slightly to 690 metres and also setting my deflection drum, adding a wind correction. I confirmed this data against my ASATS cards. We were both now observing this area with maximum concentration: nothing else mattered. I was using my rifle scope, first tuned up to increase the power to a x14 magnification, now and again going up to x20 and back down to x12, then back to x14 depending on the varying light conditions and their subtle changes. The ANA officer was using his binoculars. He adopted the same sitting position as me and would occasionally switch to using the spotting scope.

Suddenly there was movement at the target end: the likely antenna was being moved from the base, moving very slightly to the right, and with that the torso of a figure in dark clothing came into view. The ANA officer started talking excitedly to me about the figure, repeatedly asking whether I could see it. I had a good side-on view of the shoulder and head area at this point, although the lower half of the man was out of sight to me. He lay flat on his front using both elbows to support himself in this prone position and his forearms and hands were raised towards his face. The word 'optics' came from the ANA officer's mouth and I caught the glint from the side as the figure moved them away from his face. That was enough for me: a man with a radio and optics may be just as lethal as one with bullets and bombs.

The ANA officer was now getting excited, while the battle out to our front was still fierce and shit was flying everywhere down in the tight confines of the buildings and streets of the town. There were two Apache helicopters over to our right some distance away, hovering while facing towards the town and heavily engaging a compound out into the far distance right on the edge. Through our optics we could make out the dust and debris plumes being kicked up from this area as the compound was fired on by the attack helicopters.

I quickly built up my position and prepared myself to fire, while the ANA officer was observing through his binoculars and trying to tell me what the target was doing. I focused on the centre of mass of what I could see, placing the centre of my graticle pattern onto the centre of the shoulder area while the guy was in the prone position. Adjusting the magnification on my rifle scope to x18, I then adjusted my fine focus: the sight picture was once again crystal clear to me. I could plainly make out the radio handset and optics the target was using, and that was enough for me to take action.

I steadied my breathing as always and operated the trigger, slowly pulling it back for that unmistakable single magnificent sound as the round is released from the rifle on a one-way trip towards its target. As it sliced and zipped through the air, I waited and watched for impact. Then, smack! The round entered the insurgent's body with such force that on impact against the flesh it seemed to lift and twist the body slightly up off the floor and then back down again, motionless, all in a split second. Both the ANA officer and I continued watching for movement but there was none, not even any twitching of the limbs. To me it all seemed very clean and neat: an 8.59mm sleeping pill. A further few minutes of observation through our optics revealed the insurgent still lying there motionless. I chambered another round and applied my safety catch while waiting to see if anyone else would come up onto the rooftop to grab the radio and optics, but nothing.

I took my head away from the rear of my scope but kept the rifle in my shoulder and continued waiting and watching using my binoculars for about ten minutes or so, then rested the rifle butt on my left thigh. At the same time the ANA officer was still seated and observing with me through his binoculars. Eventually he removed them from his face and we just looked at each other for a moment. I remember vividly that a small smile broke through on his very weather-beaten countenance to reveal some severely stained teeth. He was saying 'Good, good, no more radio problem, good' as he got up and walked over towards a group of ANA soldiers in the far left-hand corner of the rooftop who had now joined us. I picked up my ejected round as usual, put it away and quickly scribbled down the details of the engagement in my firing log book. Packing away some of my equipment, I prepared myself to move: I had to get some water down my neck as my head was now throbbing. I looked around and out into the middle and far distances. Time for me to move on again and try to relocate as the systematic clearance of the town was moving forward and gaining ground.

Chapter Six

Patrol Base Argyll

Part IV, Operation SOND CHARA

The last few hours up on the rooftop with the ANA officer had passed by quickly for me and the onset of early evening was now rapidly approaching. It was time for me to move and try to find and link up with the remainder of the platoon. A few of the Scimitars were up at the front with the most forward call signs in contact and the other half of the platoon Scimitars were providing flank protection and picketing the routes on the outskirts of the town. The wagons had been in and out of contact for most of the day providing fire support and punching their way forward with their 30mm RARDEN cannons, taking on and destroying several insurgent bunker positions.

They made maximum use of their daytime optics and thermal sights within the turrets for observing and reporting information on to the rest of the battle group regarding insurgent activity and locations that had been PID en route. This gave them an added advantage when observing and scanning over the ground, especially in locating and identifying a number of pre-prepared insurgent bunker positions that were easily destroyed by the 30mm HE rounds slamming into them in three-round bursts, each exploding on impact in rapid succession, one round after another finding its mark. Landing in and around the small bunker positions, the rounds threw up a mixture of clumps of sodden earth, water and wood debris a few feet into the air with each strike.

As always, the insurgents were also using the irrigation ditches that covered the landscape for movement, cover from view and fire. These led to and from their pre-prepared locations and when PID by the forward Scimitars they were engaged by the 7.62mm coax-mounted machine guns firing a mixture of ball and red tracer in a seemingly continuous red line

A spotting scope and binoculars are essential pieces of equipment carried and used by the sniper on all taskings and must be just as carefully maintained as the rifle and the ammunition itself.

Pictured left is a lightweight Arctic winter snow overjacket that has been dyed and oversprayed to match autumn seasonal colours. On the right is a standard construction Ghillie vest used by British snipers, with additional frayed hessian material strips that have been dyed, oversprayed and finally weathered to give the vest a more natural appearance.

The L115 A3 pictured in the foreground is fitted with a suppressor that can be easily attached or removed from the rifle. The sniper on the ground will decide if the tactical situation or threat at the time requires him to fire with or without the suppressor; he considers the advantages or disadvantages when calculating his firing data.

A size comparison of the small-arms ammunition that can be carried by British snipers on the ground. The largest is the 8.59mm calibre L115 sniper ball rifle round; centre is the standard 7.62mm calibre L96 rifle round; the smallest is the standard 5.56mm calibre SA 80 A2 rifle round.

Pictured is a sniper pair in the prone position without any natural screen, backdrop, depth or trapped shadow to aid concealment. By paying attention to the front of the optics, in particular the colour and shape, the scope ring could be picked out by a trained and experienced eye. Therefore various methods and means of concealing the optical lenses are taught during sniper training.

The Balkans: operating as a sniper pair somewhere in a remote mountain village. Note that for our own protection the 7.62mm GPMG is carried within the pair for its reliability, stopping power and rate of fire.

The Balkans again: our observation post was the derelict building behind me which we occupied for some time while observing a possible insurgent re-supply route. Using our spotting scopes and rifle optics we were able to positively identify and trigger the insurgents. The end result was that weapons, munitions and several insurgents had been seized the night before.

Training for deployment to Iraq. Rapid bolt manipulation is a key skill practised by the sniper for multiple target engagements. For stability and accuracy the prone position is preferred, creating minimum disturbance to the sniper's firing position and sight picture and ensuring the rapid and effective take-down of multiple targets.

Iraq: grouping and zeroing L96 sniper rifles for the final confirmation of firing data. This ensures the rifles are accurately zeroed to the firer, especially when engaging targets at greater ranges. Adopting alternative firing positions is also practised for the final time, just before deployment into the city of Al Amarah.

CIMIC House pictured before fighting started against the insurgents. This building and its small compounded location, which we had to defend and fight from on a daily basis both day and night, would become our company strongpoint during the long hot summer months of 2004.

A profile is given by adopting this firing position: this was a permanent overt observation and firing position sited up on the rooftop of CIMIC House. The advantages of this position were clear fields of observation and fire for us, more or less 360° around our location.

A Snatch Land Rover after a mobile patrol in the city of Al Amarah. Very often our vehicles would take a hammering as more or less every time we left the front gate to patrol, either mounted in these vehicles or on foot, at some stage the patrol would end up in a fire-fight against the local insurgents. A fair few of these vehicles were either destroyed or severely damaged during our tour.

Pictured is a very large controlled explosion on the outskirts of the city of Al Amarah. Note the height of the electricity pylon over to the right of the seat of the explosion. This was a detonation of just some of the insurgents' vast array of available ordnance ready to use against friendly forces; found on searches in the local area, it was taken away and destroyed.

Afghanistan: shortly after arriving in our new location at Patrol Base Argyll and conducting preliminary observation from the rooftop. Arcs of observation and fire and key ranges and reference points are recorded; also the wind direction and strength and the light conditions. Equally important is study of the local pattern of life which is carefully observed.

Note that as a pair there is no attempt to conceal ourselves up on the rooftop while conducting observation of the local area. The sighting of any long-barrelled weapons or specialist optical equipment is of interest: the sighting of this equipment overtly works both ways by letting the insurgents know that we are there.

Pictured is a Chinook helicopter shortly after take-off having picked up a battlefield casualty during Operation SOND CHARA. The men and women of the MERT (Medical Emergency Response Team) often came under insurgent fire when coming in to land or during take-off. The skill and bravery of the helicopter crews and medics in the air and on the ground under contact saves many service personnel lives by stabilizing them medically, then getting them out of danger and back to Camp Bastion field hospital for further life-saving treatment.

Halfway through Operation SOND CHARA and a small smile manages to break through on my face having just met up with the remainder of the platoon on the ground: it is good to be back with them for the remainder of the operation.

A mortar crew from 2 PWRR in action with the 81mm mortar. Such crews were on constant standby to provide immediate fire support to the troops in contact fighting on the ground, able to provide HE, smoke or illumination fire missions. The MFCs (Mortar Fire Controllers) would be forward with the fighting troops and in a position of observation to be able to call in the necessary fire to neutralize a given threat.

A detailed reconnaissance of the ground by the sniper as and when possible is vital, especially when operating in a new and hostile area, in order to gain information on the terrain, the locals and habitation. Note that for this task my kit and equipment worn and carried should not draw attention to me or to my role as a sniper.

Coxy and me at a platoon rendezvous during Operation SOND CHARA catching up on recent events. The platoon is being re-supplied with ammunition for the Scimitars, 30mm and small-arms ammunition; also fuel for the wagons plus bottled water and rations.

Pictured is A Company 3 Platoon 2 PWRR commanded by Lt D (the boss) and his second-in-command Sgt M (Moggy) with personnel attached to the platoon. The unit took three casualties in the last month or so of its deployment: one individual received a GSW and two suffered shrapnel wounds to the upper torso and lower limbs from RPG fire. All three made a full recovery.

Manoeuvre Support Group (MSG) humour.

Pictured from left to right: me, Spenny, Sam, Joseph and Matt shortly after our first night engagement together. The use of our night-viewing devices and other assets proved to be a real battle winner.

A good close-up of Sgt M (Moggy) with his ageing, reliable L96 sniper rifle in the aim during an operation with the PEF (Poppy Eradication Force). Later that afternoon we both had single engagements out to a range of 880 metres, the wind and light conditions being good for a late-afternoon engagement.

Moments later while providing cover for the boss we were involved in a very lengthy engagement against the insurgents who in the later stages of the attack tried to assault our position directly. When that failed, they tried to manoeuvre around and outflank us in our position, which was also unsuccessful.

Big Sam on the Javelin system. Use of the Javelin and other assets was always a sure battle winner.

The 60mm mortar in action. In our platoon location this weapon was expertly manned by Cpl W (Stu) who earned the nickname among us as the 60mm sniper for his accuracy in putting down fire onto the insurgents when called for.

The recce platoon from 1 PWRR pictured together at the end of our tour in Afghanistan. At this point in time our comrade Bill was back in the UK, recovering in hospital from his GSW.

Pictured from left to right are Matt, myself and Ben in our final days in Patrol Base Silab at the end of our tour in Afghanistan. This was my last operational tour as a sniper; it was time for the old man to hang up his boots. As for Matt and Ben, they would be returning shortly.

My sniper propaganda corner: always, when given the opportunity within the battalion, promoting the role of sniping within the regiment. I hoped to encourage new young blood from the rifle company's latest intake of soldiers in wanting to train and become the next generation of British army snipers within our regiment, keeping our unique skills alive.

that arched slightly, following the curvature of the earth for a moment, then coming back to earth and impacting into the ground.

I double-checked that all my kit was packed away and went through the little checklist in my head: which piece of equipment was packed away where, should I need it in a hurry. It did not take long to break down and level out the pile of rubble that I had built up and around me in the area of the smashed-up pillar. I dragged and pushed the rubble around the rooftop, putting as much as I could back in its previous location. It would have been impossible for me to cover all my own ground sign but to me every little helps. Anyway, the position had been occupied by platoon strength plus in numbers at times.

Whenever and wherever possible, especially when extracting from observation posts, firing positions or any occupied or used location, it is imperative that we leave as little ground sign or information about ourselves and our intentions for the enemy as possible. They will come along and try to gain as much information about us as they can. This could be anything from patrol or formation numbers which might be gained from our footprints and direction of travel, to the depth of our prints in the mud or sand revealing whether we are travelling in light order with just belt kit or fully loaded carrying bergans or support weapons, or whether we are moving slowly with caution or pacing it out with a longer stride. This is a hugely important subject covering many things and every soldier must be aware of its implications. Everything from used camera or shaver batteries through to discarded rations must be disposed of through the correct chain. Any such items, including pieces of kit and equipment just thrown away or lost on patrol, can and will be used against us by the enemy.

The roof was now getting quite busy with ANA soldiers and some members of the ANP who I believe were going to occupy and maintain that position for the remainder of the operation and use it as their operating base. Some of the Royal Marines who were left on the rooftop and around the building at ground level were now preparing themselves to move out and link up with their own call signs that were forward and pushing through the town or out on the flanks.

I came down from the rooftop, down what remained of the rubble stairwell that was much easier to negotiate in daylight compared to earlier that morning when I more or less stumbled my way up there in complete darkness. I came down into a large room at ground level and exited the building via double doors. The glass from these doors had been blown or shot out and now covered the floor, intermixed with lumps of rubble and masonry. The walls above and to the sides of the entrance were covered in strike marks from small-arms fire. I remember at this point that my boots, trousers, body armour and the pouches on my body armour were covered with a whitish, chalky residue from the rubble up on the rooftop and as I came out of the building I was trying to rub it off, because compared to the troops outside I looked like I had been rolling around in a chalk pit all day as everyone on the ground was covered in dark mud.

In the small high-walled courtyard there were ANA soldiers, ANP and a few Royal Marines heavily laden with equipment and two dog-handlers with their dogs waiting patiently to move off and get moving forward. A Viking vehicle was parked up on the other side of the wall with its engine running. The air temperature was starting to drop, as were the light conditions. The sun was starting to go down, moving into last light, and I could physically feel the sudden change in the air temperature around me. I had been sweating like a lunatic all day just like everyone else on the ground and had not taken in much water. That factor, plus other restrictions on the body with body armour, belt kit and a helmet on your head all day, means you very quickly start to dehydrate. As the hours pass by you don't really notice it until you stop. But once the body calms down and the adrenaline goes, that uncomfortable feeling creeps up on you slowly and kicks in: feeling drained of energy with a very dry mouth, a cold clammy sweat and a pounding head. It is not good for the body which needs fluids while cooling down, with maybe a sachet or two of Dioralyte rehydration salts added to your drinking water.

It was such a relief for me to take off my helmet and rub my pounding head, and my eyes were red raw from observing through optics all day. Another telltale sign for me to get some fluids on board was my urine being bright orange and stinking. I sat on a low wall with some ANA guys and Royal Marines and started to drink some water, forcing myself to drink steadily and smashing a chocolate bar, some biscuits and paste down my neck. The CQMS was cutting around the courtyard giving out bottled water and spare rations to everyone. I started to cool down and could feel the evening chill as the wind picked up. I had my rifle, my trusty .338, resting on my lap and attached my GPMG sling to the rifle as this made it easier for me to carry while patrolling on foot, especially when trying to get over, under or across obstacles in the dark.

There were even more ANA and ANP arriving at this location and their 4x4 vehicles were starting to fill the area, parking up outside the front main entrance. I could hear in the distance the familiar sound of the old CVRT engines, spluttering and groaning as the support element of the platoon made their way towards us. I went out of the courtyard onto a muddy track to see if I could spot the approaching vehicles and saw the black exhaust smoke in the distance before I saw the actual vehicles themselves. What a sight that was: kit everywhere was hanging off the sides and everything covered in mud that had been splattered high up onto the sides, front and rear of the vehicles.

The old faithful Samson was the lead vehicle trundling along: a member of the CVRT family of vehicles, a variant crewed and used by the REME as one of the recovery wagons for the platoon. It was used when the Scimitars were having dramas with anything electrical or mechanical; for example, the tracks, running gear or engine, or even failure of the main 30mm armament or 7.62mm machine gun. Sometimes a tow out of the mud was needed when the wagons became severely bogged in and digging out wasn't an option. Pete was in command of that vehicle and was an experienced SNCO, a Recce Mech [recovery mechanic].

Following behind the Samson was the Samaritan with its big Red Cross identification panel on the top of the back of the vehicle: this was the platoon's battlefield ambulance. Fish was the driver of that vehicle and it was commanded by an SNCO medic from the RAMC who was attached to the platoon for this operation.

The vehicles made their way slowly towards my location through the deep mud, churning it up even more, and came to a halt just outside the main entrance by the outer wall surrounding the building. I went over to meet them with all my kit, hoping to hitch a ride. First I spoke to Fish: he was all excited and launched straight into telling me about his day and how the insurgents had tried to brass up his vehicle. At some point during the morning both vehicles had been targets for some insurgent small-arms fire and the area around Fish's head, in particular the driving cupola, had some strike marks with small indentations in the metal grilles and body frame of the wagon. I thought 'Jesus, how fucking lucky was Fish' as he was gassing on and then I noticed that my grab bag, which I had asked him to bring out to me from Bastion, was on the wagon and secured just above the back of his head area. On climbing up onto the wagon to talk to him more easily over the engine noise, I saw that my grab bag with some spare clothing in it had also been hit and had a few holes from the insurgent rounds, but better in my grab bag than in Fish.

We talked for a little while and I put my kit in the back of Pete's wagon for now. I hitched up with Pete and Fish so I could now go back and work with the remainder of the platoon for the rest of the operation as this was the only means of transport within the platoon that could carry a spare bloke in the back.

For the next few days I travelled in the back of the Samson which was awkward and a pain to get in and out of while wearing body armour. I had to remove some pouches just to get in and out of the tiny door at the rear of the vehicle and it was just as small in the back. My .338 was always in my hands so it would not and could not get thrown about while we were moving over uneven ground as this might affect the zero of my rifle; I constantly cradled it like a baby. There were some advantages to travelling around like this, mainly the fact that it was dryish with some protection from the weather, especially the heavy downpours of rain that often occurred both day and night. The Royal Marines were usually tabbing everywhere on foot from one objective to the next over the now sodden muddy ground, taking and clearing compounds, occupying them for a period of time and then moving on to the next.

Each vehicle had a cooker which worked if you were lucky. Believe me, when travelling in the back of Pete's wagon everything was going to work 100 per cent, especially the cooker. So we were able to have hot food and brews which was an added bonus, I thought. So many times in the past on operations and on exercises it was hard routine for a number of tactical reasons. When deployed as a sniper or in a reconnaissance role on a certain tasking, depending on the type of task and threat it may be necessary to carry out that task in hard routine. This means conducting your skills and drills at a very high level of battlefield discipline in order to minimize the risk of being compromised by the enemy.

This could mean, for example, that on an OP task there was no cooking inside the OP; you had to urinate in plastic bottles and defecate in clingfilm or plastic bags and then have to carry all your rubbish and body waste with you while on task, as well as all your kit and equipment required to complete the job. I was one of the lucky ones out on the ground. At this point during the operation it was only my boots and lower legs that got really wet and caked in mud and it was some comfort getting into the back of the vehicle after a tasking or contact. Some of the Royal Marine call signs were living in trenches and bouncing from compound to compound for this operation, constantly exposed to the harsh elements and having to live and fight in these conditions on the ground. For now I was in the back of Pete's mobile combat caravan, as I used to call it. Hot food, brews and some protection from the elements: a real winner and I thought myself very lucky.

When the vehicle halted I would get out and move not too far from it, going over off to the flank and finding a position from which I could observe and fire. I would observe out into the middle and far distances (600 to 900 metres and beyond), picking out prominent features and man-made structures on the ground, then more deliberately and cautiously scanning these points of interest, coming closer into our location to roughly 500 or 600 metres. The same method would be applied to the ground immediately surrounding our position within 300 metres or so, concentrating on the irrigation ditches, tree lines and compounds; anything I could see that was a potential insurgent firing position onto our location. We also plotted down some key ranges on a hastily-made battle card for this specific halt. For this I used an A4 piece of clear talc [high-strength polycarbonate sheet] and permanent marker pens, placing the talc over the top of my orientated map and securing it with a small bulldog clip on each corner, then plotting key ranges and reference points onto the talc.

Fish had now become my No 2, the observer, and was light role recce-trained with a good clear understanding of the basics of sniping. When we deployed out on foot Fish, for our personal protection, had to carry the GPMG and a fair amount of ready-to-use 7.62mm ammunition.

It was just after midday and our call sign complete went firm-halted on the outskirts of one of the approach routes into a town called Zarghun Kalay. This particular day would be one that Fish and I would always remember. We had gone firm on a mud track just wide enough for one vehicle. On either side of this ran an irrigation ditch paralleling the track in the direction we had just come from and continuing on towards Zarghun Kalay. The ditch on both sides was filled with deep, running muddy water which then filtered into the large muddy fields to either side of us, extending as far as the eye could see. A few compounds of various sizes broke up our view of the landscape.

Large pools of water covered these heavily-sodden fields and what looked like the odd small pile of foliage and debris was dotted around the landscape. These needed our attention: we had come across small purpose-built bunkers and trenches with poor attempts to camouflage them before, and they had been destroyed by the 30mm. So these locations and irrigation ditches were scanned and observed as a priority. My method was to look for a pattern, as just like us the insurgents would have positions that could mutually support each other with interlocking arcs of fire and so on. Or it may be an insurgent OP position, paying attention to our numbers, kit, equipment and direction of travel.

We had stopped on the mud track, both vehicles in file and closed up together so they were both covered from view from the left side of the open fields that offered good clear arcs of fire and observation to the insurgents around us. Just outside a newly-constructed and well-maintained small single-storey compound with an outer wall encircling the building and garden area was an iron gate, the main entry point leading into and through a small well-kept garden. This appeared immaculately maintained with green grass and flowers on either side of the narrow concrete pathway leading from the gate up to the main entrance of the building. The main door was a large, solid-looking brown wooden affair with a single low window to either side.

The idea was for me and Fish to try to get up onto the roof of the building so we could observe from an elevated position the general area and the intended route into Zarghun Kalay, mainly checking for obstacles to the vehicles. The remainder of the crew stayed with the vehicles with both engines running; the radios and GPMGs were manned, and other call signs from the platoon were in contact as they approached the town. In the distance the exchange of burst fire and dull explosions could be heard and above us every now and then Fast Air were patrolling. Hunting in pairs, they would cut across the sky at speed, drowning out most of the sound for a moment or two as they flew over us and away in a couple of roaring seconds, leaving just a vapour trail in their wake.

We moved cautiously as a pair, not knowing whether the building was or wasn't occupied, covering each other; me with my .338 and Fish with the GPMG. We first checked and cleared the gate before moving through it, then proceeded slowly and cautiously along the path towards the main door of the small compound building. There were no dogs or other animals in the garden as we approached the main entrance; everything appeared to be deserted. We were fucking cautious: no animals, no people, but with every step we were looking down at where we were putting our feet, looking at ground level, then waist level, then head height as we moved along; stopping, going down on one knee, scanning the ground ahead of our next move and listening, always listening.

We reached the doorway, with the small-arms fire in the distance seeming to get louder and more ferocious as longer bursts were cutting through the air. At this point I was waiting for something to suddenly go bang or a burst of fire come towards us from somewhere inside the compound walls. We got close enough to peer in through the windows. What we saw was a large open room behind that big door that was completely empty of the owner's possessions: no furniture, no clothing, no cooking utensils, not one item; only a solid red immaculate shiny tiled floor. Happy with that, it was time for us to move on. I wanted to check the rest of the garden and see what was on the other side of the compound; the garden area that could not be seen from the track where the vehicles were parked up. We were also looking for a suitable position to access the roof and gain elevation so we could observe further down the track and view the surrounding ground where the vehicles would be moving.

We moved cautiously through the garden where the foliage, now a mixture of long grass and bush, was waist-high. We'd been moving slowly along the rear wall of the compound and could see a section that looked like it had been blown: there was a large gap, and some rubble debris was all that remained of this small section of wall as we approached from the side. It appeared to have been blown into the garden rather than out and the damage looked fairly recent. We moved along the length of the wall, using it as a screen against whatever may be on the other side; stopping, kneeling, looking and listening, then moving on again towards the gap. On reaching it we both remained low, observing and listening beyond the wall. What could be seen was another compound about 300 metres away directly to our front. The ground between us and the observed compound was just open, extremely muddy and flat with a small copse of trees off to our forward left field of view at a range of about 150 metres. We could still hear the sporadic sounds of automatic and single small-arms fire coming from this general direction.

By now the pair of us were soaking wet from sweat and the fine drizzle of the day that had turned into a steady shower of constant rain. However, visibility was still okay considering the poor weather and light conditions. We were almost back to back, me with both knees firmly pressed down into the mud on the ground, my arse cheeks resting on the heels of my boots, hunched over observing through my binoculars using both hands to steady them as I scanned the open ground with my .338 next to me. I covered the front and Fish covered the rear with the GPMG. We had been observing and listening for about ten minutes and something was happening over there in the middle distance. The sounds of automatic fire and sporadic single explosions could still be heard coming from that direction and odd plumes of dirt and debris were now rising up into the afternoon sky when suddenly there was movement observed out to my front in the area of a small compound in the middle distance.

This was along the base of the compound wall coming from what appeared to be the off side, at first out of sight to us but then moving round and coming into full view, in our direct line of sight, and they were not friendly forces. I thought 'Fucking hell'; my heart began to pound and I could feel the pulsating effect of this in my right temple. I watched and squinted a little through my binoculars, not wanting to miss any detail of what I was seeing for at least several more seconds. I counted five armed insurgents as they came around into my field of view, one by one with minimal spacing between them. Immediately I brought Fish round and gave him my binoculars to confirm what the hell I was looking at as I couldn't believe it myself. Neither could Fish as he observed them moving across our front.

I then felt in an instant a surge of heat race over my entire body, my heart going into overdrive with powerful, rhythmic beats, accompanied by a slight throb in my fingertips and toes with every heartbeat. That uncomfortable very warm feeling, the immediate rush of adrenaline and nervousness, that familiar uneasy sensation swept over me, all in a few short, excitable moments.

The insurgents were moving along the base of the wall in single file, more or less directly in front of us. In my mind, my thoughts ran straight through our options (a mini combat estimate). I thought one .338, a bit of rapid bolt manipulation, and laid out two of my .338 rifle magazines ready to use; Fish with the GPMG and a few long bursts of 7.62mm initially down range, followed by deliberate short accurate fire from the gun; we'd smash them in minutes. If they then went left or right back along the wall once we opened up on them, or even came forward towards us or just went straight down to the ground for cover, we could still take them. I quickly explained my thoughts to Fish and we rapidly started to set up: there was not a minute to waste as Fish brought the GPMG round and I adjusted and extended the bipod legs on the .338 in order to fire from the prone position.

Then in an instant there was one fuck-off almighty explosion and we both threw ourselves to the ground, thinking 'Jesus, shit, shit, what the fuck was that?' and nearly physically shitting ourselves. A few moments passed and we both kept our heads down; it was one almighty big bang. After a further few seconds we looked at each other: well it wasn't us, we're still here and we both looked back up slowly towards the area of the wall where the insurgents had been and all we could see was a massive dirt cloud at ground level with a great plume of debris starting to rise up into the sky, carried on the wind. They were gone, fucking gone, all of them, and so was a large section of the wall that had been their backdrop as they moved. We watched the area for several more minutes as nothing stirred, totally amazed and probably shocked at what had just happened in front of our eyes but hey, what a cracking end result! Another call sign must have spotted them earlier than we had and had eyes on them, tracking them along to this point till they were engaged. There was no further movement at all now, just the dust and debris settling as it came back down to the ground in the continuing drizzle.

These insurgents had been dealt with in a spectacular and decisive way and would no longer be a threat to any friendly forces call signs on the ground.

So now we had to get moving from our present position and make our way back to the vehicles as time was ticking away after all that going on around us. And now some of the forward and flanking call signs were obviously in contact again with the insurgents, as even I could clearly hear the sound of small-arms fire as the two sides engaged with each other somewhere out into the distance.

Things were certainly starting to liven up again all around us and we had to rejoin the vehicles. The guys would probably get the radio message to move forward soon anyway to keep up with the reconnaissance screen as the Scimitar vehicles probed forward towards the outskirts of Zarghun Kalay and its approach routes. A few of the Scimitars were still heavily engaged in fighting the insurgents using their 30mm RARDEN cannons against a number of pre-prepared and slightly better concealed bunker positions, and insurgents using the irrigation ditches to fire from were located along the routes and their approaches. We quickly sorted out ourselves and our kit to set off rather rapidly and moved back through the garden along our proven route, still exercising the utmost caution with all our movements as a pair until we were safely back with the vehicles.

On reaching them, we both mounted up quickly. Fish climbed up onto the wagon and squeezed his long body (Fish is about 6ft 8ins tall) straight back into the small confined space of his driver's compartment and prepared the vehicle to move off. I got in the rear of the wagon and closed the heavy door behind me, sitting as close to it as possible should I need to exit the vehicle quickly, and pushed my day sack up behind me and secured it out of the way with netting. As usual, I secured my .338 rifle by laying it across my lap as I prepared myself for the move. One hand was holding the rifle down and the other was raised above my head, pressed flat against the inside roof compartment and pushing me firmly down into my seat. We moved off with a violent jolt forward as Fish put his heavy foot down on the accelerator and the vehicle spluttered into life, going forward again with another violent bump as he changed up the gears to increase the wagon's speed and get us moving.

After we had been mobile for about half an hour, even over the engine noise and the constant vibrating and grinding of the vehicle's heavy tracks over the road wheels and ground I could just make out the sounds of automatic gunfire every now and then. It was very hot and uncomfortable in the back and any soldier who has been in the back of an armoured vehicle while serving in Afghanistan or Iraq will tell you that it's no joke being a dismount while travelling from point A to point B. The lights didn't work too well either, going off and on periodically. I was being thrown around sometimes like a rag doll, clinging to my rifle as we went over the undulating ground at speed and constantly swearing to myself under my breath. I was now constantly in an uncomfortable all-over body sweat, saturating my clothing from head to toe, and with a pounding head under my helmet I tried to take warm water down just to keep myself hydrated as best I could. It was rather like being on some theme park ride that you just couldn't get off. I hated being in the back of the wagon anyway as my situational awareness was nil and I had lack of control over any given situation. The possibility of seeing something and then acting accordingly was very limited and this is where complete trust in your comrades, the vehicle crew, comes in.

Then in an instant the vehicle was abruptly brought to a violent halt. The front seemed to take a nosedive down, raising the back end up off the ground. I was thrown from my seated position by the back door down towards the front of the inside compartment and a few loose items ended up on top of me in a heap. I thought to myself while trying to get upright: 'What the fuck was that? What the fuck has just happened? There was no explosion. Has the vehicle just rolled?' I could hear Fish's and the medic's voices from my position in the back. The medic was in the commander's compartment cursing like mad and they were both trying to figure out what had just happened and trying to understand why the vehicle wouldn't move. The internal lights in the back compartment had gone out but a small amount of sunlight coming in around the commander's hatch was enough for me to see what I was doing. I managed to get myself free and upright, got my head torch out of my gear and used it.

I found my rifle, freeing it from under a well-packed medical bergan and some opened and half-full 24-hour ration boxes that had emptied out over the back of the wagon, and on getting the rifle back in my hands I quickly gave it the once-over. The safety catch was still on, the magazine was still on and so was the optical sight, with nothing damaged or broken off the optical scope body itself. Well that's a good start at least, I thought to myself.

The .338 may look like a sturdy piece of weaponry but must be treated and handled with due care and respect because the slightest knock to the rifle or optical scope body could have an effect on the zero of the weapon. This may produce a small error in your fall of shot at the target end. As the range to the target increases, so will the error, making all the difference when you are about to engage a threat or are already in contact against a hostile force. Being able to engage that threat with a good clean single fatal shot or possibly wounding or even missing the target could determine whether or not he gets to live and fight against you another day.

I managed to get myself and my rifle back up to the back door and positioned myself to try to open the single heavy door but gravity was well and truly against me. I used the door lever and pushed and pushed, just managing to get the thing open slightly. Daylight started to penetrate into the back of the wagon through the small gap that I had created in breaking the door seal. I struggled for a while against the sheer weight of the armoured door, gradually pushing it open to its full extent and finally managed to lock in the door arm, a small solid metal bar attached to the door that hooks into a hole on the outside, preventing it slamming or shutting on someone while the vehicle is static.

I managed to climb out, getting my arse out of the rear of the vehicle, very glad to have both feet firmly back on solid ground. I quickly sorted out both myself and my kit. The ground around us was very flat and muddy and we were surrounded by ploughed fields edged with irrigation ditches. Small deciduous tree lines randomly spaced along the tops of the irrigation ditches and small man-made earth mounds or bun line features broke up the field of view over the surrounding terrain.

The mud track on which we had just been travelling continued over a mud and clay man-made culvert-type feature that covered a prominent irrigation ditch. The fast-flowing water looked deep enough to cause a real problem to any soldier in his fighting order should he fall in, and it was about 3 to 4 metres wide. The mud track continued on towards a small T-junction roughly 15 metres forward from the stricken vehicle and beyond that on the other side of the track was an open flat area of ground leading towards a well-constructed high compound wall. Some small adjacent buildings and walls came off the main wall, forming no pattern, just a habitable solid structure to our front. This was the outskirts of the town and over to the right between 300 and just over 600 metres away were more similar-looking compounds and man-made structures. We were now at the location marked on the map as Yellow 28.

As I walked from the rear of the vehicle towards the front it was still raining slightly and now very cloudy with little wind and sunlight due to the low cloud cover. I came round to find Fish at the front of the vehicle, kneeling down and inspecting the damage, and he explained what had just happened. We had been moving towards Zarghun Kalay on a straight narrow mud track roughly 500 metres from our last halt where we had dismounted and been in the compound garden. After moving about 200 metres up the track towards this T-junction, the vehicle had come under small-arms fire from the left-hand side of the track. The insurgents were fucking close; their rounds were pinging off the side of the wagon and hitting the ground in front. Fish said:

We thought fuck this shit, so we both dropped down into our compartments and closed our hatches down for our own protection. Our vision and situation awareness was cut right down and I hit the culvert slightly wrong as I tried to manoeuvre over it to get the vehicle into some sort of cover over in this area and it bloody gave way underneath to the front end. Thank fuck we didn't roll over into the water or we might all have been goners.

Looking at the water myself, observing the speed at which it was flowing and not being able to calculate the depth, I was just relieved to be out of the wagon and back on firm ground.

There was no way that Fish could get the vehicle out of this little predicament without assistance and a message (SITREP) was sent via the platoon's radio net for some help in recovery of the vehicle as the other one would not be man enough to drag it out alone. While we were waiting there were call signs forward right of our position and to our rear right. I was observing these through my rifle scope and binoculars and had good eyes onto the other friendly call signs in our area. I could see some troops stacking up and then launching themselves towards and into the area of the compounds. All that could be heard was the sound of exchanges of small-arms fire and every now and then small explosions – possibly insurgent RPG, UGL or grenades going off – with small clouds of debris rising up from the ground followed by their dispersal by the wind.

Our immediate personal security now needed some attention as we were going to be static here, waiting for some time in this vulnerable location. Time to start thinking, sorting out our position and having a defensive plan should the insurgents attack us from any direction. We started by standing to, alert and ready to fight, waiting and watching and listening for the recovery crew to turn up. We were very thin in numbers on the ground. Using my rifle scope I conducted a quick scan of the surrounding area – a 360 all around us – and pinging, looking for possible firing points and observing for any dickers (a slang term used for insurgent observers) in our immediate vicinity. They were usually about just prior to and during an engagement or incident, always watching our skills and drills or how we reacted to a particular threat launched against us.

The insurgents used mobile phones to communicate with each other, passing on their knowledge of us in their area, even if we were just moving through. A quick verbal brief would alert their counterparts in another location, or direct communication with the local insurgent area commander would quickly inform him that coalition forces were in his area of responsibility. Having received the most basic information about us – strength in numbers; on foot or mobile; kit and equipment; weapon capabilities – and our location and activity, he would ask that age-old question: 'Can we take them on?' Then sure enough in a very short period of time we could be under some form of enemy attack.

Around us the fighting was starting to intensify and seemed be moving along out in the middle distance onto our left flank. We were in a basic form of all-round defence, closely spaced with only a few metres between each of us as we did not have the manpower to make up pairs and cover more ground. Waiting, watching and listening, I felt like General Custer kneeling down in the middle of it all and just waiting for something to happen.

Sure enough, someone had noticed us and our awkward position. I had spotted him moving around directly to our front in the area of the compound wall and alleyways across from the small T-junction and sure enough, he was on and off his mobile phone constantly. By my own perception of the situation and my own previous experience of similar situations, what I was observing through my rifle scope, judging from his age, body language and overall manner, was that the guy was up to no good and something needed to be done about it.

Thoughts raced through my mind and I looked more closely at what was going on before deciding what to do. Maybe in another time and place I would have engaged him and taken him out without any hesitation but this time I decided to fire a warning shot instead that would see him off. I was still in a comfortable kneeling position and he was at less than 150 metres. I adjusted the zoom control on my scope from x12 magnification down to x5, refocused my sight picture, found a point of aim to align with on the wall, and with a final tweak of my focus brought my sight picture into absolute clarity. I warned the lads verbally that I was going to fire, and aimed just high left offset from the left-hand side of the guy's head. Meanwhile, he was still standing and observing us.

I held that sight picture for a moment or two, then operated the trigger slowly and gently and a round was released: a cracking sharp snap as the round was sent on its short journey towards and into the wall. I maintained the sight picture on my point of aim and more or less instantly a small brownish clump of the wall came off as the round impacted just where I wanted. The man must have almost shit himself in that split second of impact just above his head, as he jumped out of his skin and looked up towards the point of impact. In a flash he took to his heels, headed back down the narrow alleyway by the wall and disappeared out of sight.

The fine drizzle was now turning into more persistent rainfall and the natural light conditions were fading as dark rain clouds started to fill the sky, moving across the sun very slowly and blocking it from view for a few moments that felt like ages to me. I was standing next to the rear of the vehicle with Fish, who was still trying to figure out how to extricate the wagon. At that instant, a large shallow puddle at our feet started spitting up droplets of water that splashed over our trouser bottoms and a low zip, zip, zipping sound filled the air as rounds impacted into the water and surrounding soft mud. This remains as clear now in my head as at the actual event. Watching this, I was momentarily fascinated by the rounds penetrating the puddle of water. Then reality hit home. 'Get a grip of yourself, you lizard,' I thought to myself, 'What the fuck are you doing?' And as we looked at each other, the fuckers were firing at us. We moved to the other side of the wagon and started looking for the firing points. The blokes were in the prone and kneeling positions, looking and scanning the ground using their optics and Mark I Eyeballs.

Suddenly two vehicles, a Scimitar and a Spartan, came into sight and were heading towards our location at speed. Dark smoke billowed out from both vehicles' exhausts as they trundled through the mud, bobbing up and down, with the vehicle antennas whipping violently back and forth and looking like they might snap. That familiar sound of the CVRT engines filled the air, growing louder and louder as they drew closer. Meanwhile, more rounds were starting to come in; now a mixture of burst and single shot was landing around the wagon and the area of the junction.

I looked back towards the area of the wall and the alleyways to see the male who had received my warning shot was there once again; blatantly standing out in the open in full view to us with the wall as his backdrop and just staring, watching us. This time, however, he had a small boy with him. The lad was possibly 6 to 10 years old and the man appeared to be restraining him with his free hand on the back of the boy's neck and shoulders. When the man moved, the boy unwillingly moved with him: the adult male was forcing him around and held the boy firmly by the upper arms or shoulders and as soon as they stopped moving the boy was restrained again, held in front of the man.

I observed them both, studying them in detail through my scope with the zoom control back up to x12 magnification. Their facial features filled the scope picture and every detail of their weathered and dirty faces, even down to their brownish, stained uneven teeth and the facial hair of the adult male could be seen with great clarity. I observed the man from top to bottom and watched him for several long minutes: yes, he was watching us and maybe telling someone of our presence on the phone which every now and then he removed from a side pocket.

I decided to let it go again and without firing another warning shot. A lot of thoughts went through my head. I could have very easily taken out the adult male with a clean single head shot but I didn't because of the child; not questioning my own skill in the placement of such a shot but the possible effect on the child of witnessing such a thing. To this day I do not regret my decision, which is explained below.

Later in the day when all the ground fighting was over in our area and just before we moved off literally at last light, I remember seeing them again: the middle-aged male and the boy moving quickly along a mud track in the middle distance just as the sun was disappearing over the horizon. It appeared that the boy was still being dragged along by the arm. Then they both disappeared out of sight to me. In my mind when looking back on that particular situation (which I have thought about many, many times), I have pondered their circumstances over and over again. Were they just simply farmers, a father and son concerned at what was happening around them? Or was the man an insurgent dicker using the child as a human shield? I have seen the latter before, both in Afghanistan and Iraq, and my thoughts inclined in that direction.

Some people out there might have taken the shot without hesitation and I could have done so very easily; the range and wind didn't even enter into it. But then what of the small boy? I can just about still see him in my mind's eye: his distraught face and body stance so anxious and pathetic. What if I had fired and taken the man out, the boy then witnessing the traumatic event right before his very eyes? Without doubt this would affect him. There is already enough hardship and misery for the people here and this could be quite enough to turn him and make an everlasting impression on his young life, creating great hatred and a desire for revenge in the young boy against the foreign infidels. Yet another highly-motivated hate-driven young recruit for the Taliban, and revenge is a very good recruitment tool.

Now the two vehicles came into our location, the Scimitar facing the direction of the main threat. I got the other vehicle to park so that including our stuck one we formed a horseshoe effect (a small ring of steel was the idea), so we could use the wagons as a screen for cover, protection from incoming rounds, and a firing position. Fish went over to L/Cpl H, better known as Wild Bill, and back-briefed him on what had happened to his vehicle. Meanwhile Carl, who was the gunner in the Scimitar, popped his head up. I climbed up and briefed him as to what was going on and gave him an area to look into using the vehicle's optics and cover.

Events around us were starting to intensify even further. Looking out from our position into the middle and far distances we could see the call signs in contact, engaging with the insurgents over by some large compounds. In the distance two Apache gunships were hovering slightly offset from one another; they were working and engaging something together. Alby P came over and got down next to me with his Minimi LMG, while Bill and Fish were going around the wagon trying to figure out how to drag the vehicle back out to get it free and both were moving around in the mud on their hands and knees.

The fire directed at us now increased even more with rounds pinging everywhere around our small location. But it was hard to identify the firing positions out in the middle distance in and among the irrigation ditches and natural folds in the ground. We were starting to actually get pinned down by accurate fire that began to limit our movement.

Everyone was now either on their belt buckles in the mud or in the kneeling position using the sides of the vehicles as cover to fire around or over as and when the insurgent firing positions were spotted with target indications then being shouted. Such positions were now being pinged, located because the deteriorating light conditions in the afternoon sky enabled the men to pick out the insurgents' muzzle flash a little more easily and then engage over in the middle distance.

Luckily enough this was back down in the direction that Alby and I were now able to observe over. We were both trying to cover back down the track where the vehicles had just come up in single file and we could also observe the forward edge of the elevated area of the rooftop on top of a small compound. Our field of view was large and we could just make out some figures – four of them dressed in dark clothing – coming from the direction of a compounded area on our far left. They were making their way across the muddy fields, cutting across the open ground in small bounds and keeping a low profile using the irrigation ditches and coming in and out of cover. We could also see friendly forces and just make out the assault troops entering the compounded area in the direction these figures had just come from. Friendly forces must have flushed them out while fighting through and clearing them.

The dark-clad figures were making their escape across the fields, moving in the direction of the small compound where we had previously been in the garden. To reach this they would have to continue to cross open ground and would then come out directly to our immediate front at a range of about 300 metres, maybe slightly less. Their route would entail jumping across a wide irrigation ditch filled to overflowing with fast-flowing water, breaking cover from the ditch and then crossing the mud track right in front of us just to be able to reach the security of the compound. From their direction of travel, this seemed to be their course of action.

By the small metalled gate of the compound was a low surrounding wall across from the overflowing irrigation ditch and muddy track. There was a large tree, which as you looked at it head-on from our location had its main trunk coming up to a metre or so from the ground, then split into two prominent sections of about 20 feet in height topped with green foliage. Through this conspicuous V-shaped split in the trunk I could see straight through from our position while observing with my rifle scope.

The insurgents were trying to manoeuvre around us, and over to the right Carl had identified a position that was engaging us and started to smash some 30mm down towards them. With each 30mm round fired, the vehicle rocked back and forth slightly with a small amount of smoke emerging from the end of the barrel for a split second. After each firing the empty brass cases were being ejected by the side of the gun barrel, dropping onto the front vehicle decks and bouncing off the decking into the mud beside the track.

At this point we really had to shout to each other just to pass on vital information from our observations. I adjusted my position slightly so I was directly aligned head-on to the area with a view of the irrigation ditch and compound. The lead figure, who must have just identified us, now started firing erratically. They were in and out of cover like jack rabbits, going to ground and then coming up somewhere else in the field. They had now covered some distance and were only what seemed like a few metres from crossing the irrigation ditch and crossing over to the sanctuary of the small compound.

Alby and I got up and moved again, going back onto the mud track and adopting the kneeling position once more. We were both now looking straight down the track and the field of view and fire was not going to get any better than this. Alby was kneeling on my right side, the plan being to get them as they crossed the open track. He started to fire his Minimi in short bursts down towards the direction of the insurgents who were now in the irrigation ditch. Link and empty cases were spat out of the Minimi gun with each burst of fire and dropped to the ground where he knelt. Suddenly from behind us Bill came running in, took up a kneeling position with his A2 right beside me to my left and started firing his rifle; his empty cases were hitting me and bouncing off my shoulder and the side of my helmet. Incoming rounds were going everywhere, striking into the mud, hitting the vehicles and ricocheting off.

I was sandwiched in between the two men who were firing single shot and burst fire and a mixture of ball and red tracer was zipping through the air at waist height down the track towards the insurgents. We were almost shoulder to shoulder and the noise of the rapid fire and the 30mm RARDEN cannon together with everything else was mind-numbing. Some of the rounds fell short and were striking into the ground, spitting up small pieces of mud and foliage along the bun line of the irrigation ditch. In some places there was just thick green foliage: we could barely make out the insurgents' muzzle flash at this range and they seemed really close.

I was scanning the ground left and right slowly up and down the irrigation ditch where they went out of sight and waiting for them to break cover to cross the track. Taking a quick moment away from my scope I looked around and tried to take in everything else that was going on around us. Back to my scope, I was looking for any movement, muzzle flash, or anything that might get me onto a target. Observing along the ditch, the foliage was thick in places and probably made a good screen for them. An insurgent abruptly appeared, seeming to come up and break cover from the ditch and dart across the track towards the tree in an instant and he went out of sight again, using the tree as a screen. We shouted to each other that this had just happened and were then waiting for the others to cross.

I took the small area of ground around the tree to observe, while Bill and Alby took the main track. I was completely focused on this tree area.

It filled my scope and I could just make out some sort of movement behind the tree but not enough to engage. He must be either lying down with his body aligned behind the tree or standing using the tree in a similar fashion; either way he was going to be stuck there and waiting for his comrades to cross over. But he had to get up or make a move at some point, I thought to myself, so I was completely focused on the V-gap in the tree and waiting for him to make that move. Both Bill and Alby were still engaging the ditch area directly opposite the tree. The insurgents must still be in there somewhere and preparing to cross or just waiting for the right moment.

The V-gap of the tree had light behind it and suddenly that gap started to darken. I adjusted my fine focus and could make out the outline of a figure that looked to be facing back across the mud track, slightly offset to me at an angle, and making some sort of low signalling gesture to the others to cross. I could distinguish the lighter colour of his fleshy hand compared to his dark clothing. Tightening my whole firing position slightly I then placed the centre of my crosshairs onto the centre of observed mass and adjusted my focus slowly for the final time until the sight picture was the best I could get.

Now ready and poised to fire, I had already removed my safety catch and was just about to operate the trigger. For a short moment I hesitated and rested my finger lightly on the trigger as the guy quickly moved and in an instant he had turned round to face towards us through the gap in the twin tree trunks. Just then Bill and Alby started to rapid fire; his comrades must now be crossing the road. I at the same time was poised observing through my rifle scope, then saw the insurgent and his weapon as he filled my sight picture. It was so clear I could see that he had an AK variant with a UGL attached to it and even noticed a bright multi-coloured local patterned sling attached to the front of his rifle. Instinctively I brought my point of aim down just slightly and it was now on the upper half of the torso.

I held that position for a further second, focusing with intense concentration. Increasing the pressure in my trigger finger I squeezed slowly, not wanting to snatch the trigger, pulling it back with the tip of my right forefinger until that familiar sound again filled my ears. The subsequent vibration of this action was felt in my right shoulder and cheek as the round was released and sent on its one-way journey to the target. It impacted into the insurgent's body almost instantly, penetrating with some force and I saw what can only be described as a puff of brightly-coloured red mist come off from a small area of the target's upper torso. The body just dropped to the ground in a heap at the base of the tree, exposing the lower half and legs. As always I continued watching for a while but there was no movement, not even the slightest twitching from an outstretched limb. I broke from my firing position, took a moment to wipe my eyes and face free from a mixture of sweat and rain and went back to observing through my rifle scope.

Seconds passed and the noise from all the firing was deafening but then out of the corner of my eye I saw what appeared to be Bill's lower legs and boots leave the ground simultaneously as if an invisible and violent force had just picked him up and thrown him backwards from where he had been kneeling just before while firing his rifle. A few moments elapsed, with insurgent rounds coming in and landing all around us, seeming to come from a slightly different direction over on the left flank. Rounds were outgoing from our location but it seemed from all the incoming fire onto our position that we were nearly surrounded. Noise, smoke and organized chaos reigned in the mud at that time.

I took a quick scan around to my left and right and tried to figure out what had just happened. It had not yet registered in my mind or been noticed by the others who were facing out either in the kneeling or prone position and still firing their respective weapons. Alby was on his knees caked in mud frantically trying to reload his Minimi with a fresh belt of 5.56mm, the 30mm cannon was still firing and only stopping momentarily to be reloaded, and two of the Royal Marines who were travelling in the back of one of the vehicles that came to help us were firing their UGLs and rifles, working as a pair and giving out target indications. Simon, a corporal in the REME who was one of our recce mechanics, was on his belly covered in mud from head to toe and crawling around firing his A2 rifle. By now the threat from down the track had been dealt with and the dead lay where they had fallen around the irrigation ditch and the base of the tree.

Moments later we looked back and Bill was lying flat on his back with one arm outstretched and the other up and slightly bent behind the back of his head. The colour had completely drained from his face and he was trying to speak. I scrambled over to him on my knees more or less on top of him and could just about hear him saying: 'Monty I've been shot, Monty I've been shot.' I could hardly hear a thing with the din going on around us and couldn't see any blood or visible signs of physical trauma to his body. He had definitely been hit somewhere, but where the fuck? I pulled him out of a puddle and undid his body armour, turning him over slightly to left and right so I could see his sides and back, looking for the entry or exit point as it was not on his front. Then there it was: a small clean entry point to the side of his trunk in the top half of his chest, slowly oozing what looked like thin, watery blood.

Alby dropped down next to me on his knees. We both looked at Bill and then at each other. Bill was conscious which was good, fucking good. We were both talking to him, constantly reassuring him and getting our field dressings out. He even managed a small smile when we nervously took the piss out of him, trying to keep him awake and alert and aware of what was going on around him. He gave it back as best he could, considering the pain he must have been in; rambling on about how could they have missed my big arse and hit the little guy? Bill was now going further into shock. There did not appear to be much blood loss in the area of the wound; however, it is what you can't see going on inside the body that can cause mega dramas if not quickly identified.

I know Alby and I were both trying hard to stay fucking calm and focused on what we were doing for Bill and putting a brave face on things. We were still receiving a lot of incoming small-arms fire at this stage. Alby started getting the rest of his med kit out and stayed with Bill while I went over and got the medic who was manning the radio out of the wagon. He grabbed his kit – the oversized medical bergan – and down on all fours he made his way through the mud over to Bill and set to, working his magic as best he could given our current situation.

The next task was to get a 9-liner (a MEDEVAC request) out over the radio but it was not that simple: we were still under contact so we couldn't call the helicopter straight into this location or even secure an EHLS 100 metres away from our position as call signs were in contact all around us and it looked like we were possibly surrounded as fire was coming in from all directions. As soon as the information was collected, the message was sent out to the rest of the platoon over our radio net, the other net being busy dealing with other serious incidents including another CASEVAC.

Communications were bad and Steve, the platoon 2IC, turned up sharpish in his vehicle. I ran out to meet him and have a face-to-face, briefing him on what had just happened and what we were going to do about the extraction plan for Bill. We had to get him out.

Everyone was either on their belt buckles in the prone position or kneeling and firing. Bill was conscious and being treated by the medic with Alby helping him. Both of them were on their knees hunched over Bill, the area around them starting to fill with blood-stained dressings and discarded packaging from the swabs and other medical items that were being used on Bill, who was growing weaker. The persistent rain had now turned back into a really fine drizzle.

We were both coming up with ideas and looking at the map for a possible EHLS for Bill's extraction that was not in contact and not too far away from this location. Steve was going to have to do a quick route recce for the possible EHLS, make sure it was suitable and then come back and get Bill. After a quick map estimate and tying down a plan, Steve's vehicle powered off in the direction of the potential EHLS and out of sight, back down the single track spitting mud high up into the air from the rear.

I got back into the centre of our small location and had to reposition the Scimitar that was still mobile so it could shield us from some of the incoming fire that had now changed direction. The Scimitar was moved right next to Bill so the medic could hopefully do his work completely shielded from incoming fire from the left flank. Time passed by and it was growing late in the afternoon. Bill was hanging in there and still conscious but gradually getting weaker. Every so often as I moved around the position I would pop over to see how he was getting on and briefed him, the medic and Alby on any information as and when it came over the radio. It seemed we were just waiting and waiting and holding our small position but we were all working together, every one of us. All we could do was wait but at the same time keep the insurgents at bay and look after Bill. The insurgents had started to move round and try to get behind us: they must have worked out what we were using the vehicle for, the fact that we had a man down and that our numbers on the ground were very thin in this location.

Ammunition of all types was getting low and had to be checked and redistributed among our group. On some of the weapon systems stoppages started to occur and in one or two cases caused some major dramas. An LMG and a rifle that I couldn't sort out became non-functional and were put on the back of the wagon as there was no time to fuck about. We were getting into a little routine by now: every so often I would go round for a quick head check, a face-to-face with the guys, an ammunition check and redistribution of any ammunition I had collected. I briefed the guys as soon as I had any information come in over the radio which was now manned by Carl; he was keeping me informed of events. Bill was stable for now and still just conscious which was good but the medic wanted him extracted as soon as possible as his condition was slowly starting to deteriorate. It was still raining and the air and ground temperature had dropped, which we could all feel through our wet clothing. Alby now had to use Bill's rifle, and I had stripped Bill's kit of his ammunition and pistol. Alby was providing immediate close protection while the medic tended to Bill and tried to keep him as warm and dry as possible.

A lone Scimitar now came into our field of view and the sound of the struggling engine and black exhaust smoke followed the vehicle with every slight twist and turn as it moved at speed, bouncing over the uneven road and splashing up mud and water as it went along. It stopped just short of our position and I could see that it was the boss Capt B (MC). I told Carl and the medic that I would be back as soon as possible, left my sniper rifle with Carl and took up my SA 80 (A2).

I got to the corner of Fish's stricken vehicle, peered around and looked over to where the boss was climbing down out of his turret and adopting a knee beside the vehicle track and road wheels. I told Fish to cover me and waited for a lull in the firing which had calmed down a little and was now more sporadic. Then I took several deep breaths and went for it. I got up and ran as fast and hard as I physically could, trying at first to hard-target down the track towards the boss. My boots were covered in mud and seemed to get heavier as I went along; my body armour seemed to bounce a little and felt restricting; and after 50 metres or so of hard-targeting I was chinned and breathless. I just got lazy, made a direct line for the boss and went for the home run. But my concentration on trying to breathe and move fast was broken by a few short bursts of automatic rifle fire suddenly zipping into the side of the mud bank along which I was running. It startled me, scaring the shit out of me, finding a mixture of mud and water spitting up off the ground around me and I just ran harder as it seemed to be chasing me.

I got to the boss, who was in the kneeling position by the back of his vehicle, close in to the side of the wagon's track. As soon as I reached him I was on my knees trying to get my breath back for several moments. I quickly back-briefed him on the sequence of events throughout the day leading up to Bill's GSW, his condition, what we planned to do about it and how we were going to recover our vehicle and get everyone back once Bill had been CASEVACed. We talked very briefly and then he back-briefed me on what had been happening to other call signs out on the ground that day and that everyone at one point or another during the day was having to deal with some really serious situations. The boss was leading and commanding the remainder of the platoon and during the advance the lead reconnaissance screen had come up against severe opposition from the insurgents, having engaged some fixed bunker locations that had MG and either RPG or mortar support as this is what nearly took out Cpl P's wagon.

It was only as my body was calming down and I was taking in the information that the boss was telling me that I noticed he had a deep laceration under his eye to the fleshy part of his cheek. He quickly explained what had happened. A near miss: either a round literally just passing by and slicing his cheek open as it passed by his head, or a ricochet hitting and coming off the support bar for the vehicle's ECM metal support frame. Indeed, there was an indentation in one of the support arms of the frame itself.

Well we had a plan and the boss had to get back up to the front where the battle was still raging between friendly forces and the insurgents. Here Jimmy's call sign had been previously heavily involved in some engagements against the insurgents and Jimmy had received a flesh wound to the upper arm and shoulder area from either mortar or RPG airburst fire around the vehicles while engaging some PID bunkers. They had to continue pushing forward providing a recce screen and locating and destroying any such positions or threats that they came across.

The boss mounted back up and into the turret and his vehicle was away almost immediately, as soon as his headset was back on his head, manoeuvring off in the direction he had just come from. I got myself back over to Bill and the medic. Bill was still just about conscious and I reassured him that what we had planned was still happening and that we would get him away as soon as possible. It was clear we really needed to get him away as his condition was deteriorating even further. The medic and I were very aware of the time elapsed since the time of wounding: by now just over forty-five minutes had passed and we were cutting it close.

I went back to Carl's wagon and climbed up on top of the vehicle to speak to him, keeping myself low on the turret, got my .338 back and checked whether any messages had come in over the radio from Steve's call sign. The afternoon was speeding by and in maybe just over an hour it would be last light. Then from atop the vehicle I could make out in the middle distance another coming back, heading in this direction and sure enough it was Steve's call sign. A sense of relief went through my whole body as I was thinking: 'At last, thank fuck for that.' We could now finally get Bill out of there and extracted back to where he belonged in Camp Bastion field hospital.

The insurgent small-arms fire had calmed down but would occasionally flare up again, becoming heavy and sporadic at times and then suddenly stop. The men were in a watch–and–shoot phase, so to speak: we had to conserve ammunition, the light conditions were starting to fade and target identification out in the far and middle distances was becoming hard. Plus we might need all our remaining ammunition for a possibly long night because we still had to recover Fish's vehicle.

I jumped down off the vehicle, ran over to the medic and told them Steve was on his way back and that he would be here in figures five [five minutes]. Straight away the medic and Alby started to prepare Bill so that he could be moved, making him a little more comfortable on the lightweight stretcher ready to be manhandled into the back of Steve's wagon. I made my way over to Steve's vehicle once he came in and we had a face-to-face brief; all was good to go. A suitable EHLS had been found, another call sign from the platoon was securing and marking the site, and a helicopter was already en route here from Camp Bastion. Bill was put on the back of the wagon with the medic beside him, sorting out his dressing and constantly reassuring him. I can still remember seeing Bill there, trying to sit slightly upright on the stretcher and a small smile trying to break through on his face. We smiled back and waved to him, saying that we would see him very soon and in a joking manner Alby told him to stop being a lightweight and man up and on hearing this, Bill just stuck his middle finger up at us, trying to smile. We closed the back door of the vehicle, moved away and signalled to Steve up in the commander cupola that he was good to go. The wagon moved off towards the EHLS back along the track and disappeared out of sight.

I called Fish and Simon in to me, our next priority being to sort out the vehicle. Si had a plan and briefed me, and shortly afterwards he and Fish set to work getting the tow chains out of the Samson and some other heavy items of recovery equipment that were needed to do the job. The insurgent fire had almost stopped but as Si and Fish were moving about setting up and laying out the tow chains, the insurgents were taking pot shots at them from over to our rear right and we could just make out their muzzle flashes along the now darkening areas of shadow running along the tops of the irrigation ditches. The natural light was fading fast, last light would soon be upon us and we didn't want to be still trying to recover the vehicle in the dark.

The remainder of the men were still in their firing positions and due to the failing light conditions I brought everyone in closer, creating a much tighter all-round defensive position that we could protect in the darkness. It would also be easier to maintain command and control, still be able to provide cover for Fish and Si while they worked on recovery of the vehicle, and finally for us all to then be able to mount up and move off. However, as soon as the insurgents decided to engage us again their firing points were soon spotted, a quick fire control order was given out and the response was to return a rapid rate of fire in that direction for a good few seconds, then observe the general area and wait for the next exchange of fire. I got a sandbag out of the rear of Fish's wagon and went back to the position where Bill had received his treatment from the medic and cleaned up the site, putting all the packaging and used dressings into the bag along with the other medical rubbish that was lying on the ground. I then bagged and tagged the remainder of Bill's kit in a canoe bag and put it in the rear of the wagon.

After some fifteen to twenty minutes of pure hard graft by Si and Fish they managed to hook both the vehicles up and eventually after some interruptions from the insurgents Si managed to pull Fish's vehicle free and drag it back onto the mud track. Suddenly we could hear the noise of engines and vehicle tracks going over the uneven ground and through the mud, the rollers and rear idlers that drove the track making a constant grinding noise as the vehicles moved along. It was the familiar sound of the Vikings in all their glory, hurtling down the track towards our location. This was truly a welcome sight for the lads, instant morale, with everyone getting up onto one knee to see the sight if they were not already kneeling or observing from behind the sides of the wagon.

An absolute sight for sore eyes: the real cavalry and Royal Marine call signs of the Viking vehicles as they came hurtling down the track in single file. The lead vehicle came right into our position spitting up mud and water from the back of its tracks and slammed to a halt with the vehicle cab rocking violently backwards and forwards. I ran over to the front and made contact with the vehicle commander. We had a quick face-to-face, he asked if they could help and I explained that our man had been CASEVACed out and that we were nearly done and good to go. But I also pointed out and briefed him towards the direction of where we were still receiving some trouble from the insurgents. Straight away they launched themselves forward, crossing the open muddy fields moving in single file, getting themselves into extended line and heading in the direction of our last known identified insurgent firing positions and off towards the sunset, disappearing out of sight to us.

Each and every one of us was well and truly chinned by now, having been running about for most of the afternoon and it was all beginning to catch up with us. But we still had to get back to PB Argyll and regroup and redeploy as the operation was still in full momentum. The blokes that were with me that day had dug out a real blinding effort; all of them a mixture of soldiers and Royal Marines from various groupings and with a wide range of skills, all operating together to make it work that day as best we could, just like everyone else out on the ground.

Finally we were able to mount back up into the vehicles after a last quick check of our location just to make sure we had not left anyone or anything behind before we moved off back towards PB Argyll. We set off slowly and cautiously, slipping away into the darkness in single file. On arrival we parked the vehicles up in file along the outer wall of the PB in the area just outside the front gate as the location was full. We could hear engines, generators, and the noise of vehicles, men and equipment coming and going on taskings.

The operation and the fighting did not stop just because it was dark. It was now in full swing and the operations room and medical bay were a hive of activity with people moving about in the mud and darkness wearing their head torches. All you could notice were lots of little lights moving around in the darkness just like small fireflies buzzing around the inside of the base. The whole area around and inside the PB was just a quagmire of deep mud as the drizzle still continued to fall.

We secured and left the wagons backed up, parked against the outside wall of the base close to the ANA/ANP gate sentry location which was manned 24/7 and was the main entrance to this location. We walked into the patrol base and met up with some other members from the platoon: we all looked in a bit of a state but the blokes were glad to meet up once again and under slightly better circumstances. They soon started to talk among themselves, catching up on recent events from the last few days. However, the general mood and atmosphere among them was not what it would usually be at a moment like this when we all met up. We were all just a little confused as to why the boss had wanted to gather the whole platoon in, extracting all our call signs off the ground. I think he was in the operations room receiving a back-brief on the day's events and a SITREP on Bill's condition.

We were all closely gathered together in a group and had been waiting for a short while standing in the mud and drizzle and then the boss appeared as if from nowhere, coming out of the darkness wearing his head torch and making his way into the centre of the platoon group. The bright beams of light from everyone's head torches shone onto the boss's face and body and illuminated him in the darkness. He was not smiling as he usually would when he was among or in front of the platoon and there was an uneasy, unnerving moment or two of complete silence from him and from all who were stood around him. Someone abruptly broke the silence and hollered: 'What's up? What's going on then, boss?' Capt B (MC) then very calmly and solemnly explained to everyone that due to the type of injury Bill had sustained earlier in the day, he had not survived his injury. That announcement was followed by total silence; instant shock from everyone while their minds tried to process the information just received.

This was not the first time that I had received news like this about a comrade and there is no easy way to explain how it feels. For a moment in time your brain tries to assimilate the information in a logical manner, taking a second or two to fully understand, to comprehend, to absorb what has just been said to you. Then follows a feeling of sheer emptiness, nothing registering in the brain, followed by doubt leading to shock when the mind finally fully grasps the information: your mate, your friend, he is gone. An almost physical feeling of wanting to be sick, a dull gut-wrenching ache, sweeps over your whole body in an instant, almost taking over. The feelings of sorrow and loss are then succeeded by feelings of anger; all this in such a short space of time.

Sometimes, despite all the training and experience in the world and despite how tough you think you are, there are some things that they cannot prepare or train you to deal with. Sometimes simply grabbing hold of a grief-stricken comrade and holding him for a few moments can make all the difference in the world: he is not alone. None of us are alone in our grief and together we would get through this and become even stronger and more focused on the task at hand.

A painful half-hour or so slowly passed by. The rain continued to fall onto the already waterlogged ground, collecting in deep muddy puddles that were everywhere, but that didn't matter much to men of the platoon at that time: they just got on with their work, prepping the vehicles to go back out. We still had a job to get on with like the rest of our comrades who were out on the ground and must get ready to redeploy back out. Then, just as suddenly as before, the boss appeared again in front of me and Steve and told us that he wanted to re-brief the platoon. So Steve and I gathered the blokes back together around the boss but this time he looked more focused and there was something in his voice. All the men gathered round and once more that night the boss back-briefed them but this time he explained that Bill was still with us. There had been a big foul-up with the passage of information, Bill was very much alive and was in Camp Bastion field hospital, and once he had been patched up and properly prepared, would be on his way back to a hospital in the UK. To put it simply the platoon were all very relieved to hear that news and totally surprised by the turnaround and outcome of events that night.

The platoon deployed back out onto the ground for the remainder of Operation SOND CHARA and, like everyone else during that time, just soldiered on tirelessly with the operation to make it successful. The main British forces on the ground consisted of the Royal Marines of 42 Commando and their Royal Artillery and Royal Engineer support groupings, C Company 2 PWRR and B Company 1 Rifles, also A Squadron and 1 Troop C SQN from the QDG.

The platoon had a fair few more engagements as we advanced to contact against the insurgents, as more ground was taken and cleared towards the closing stages of the operation, and gradually as the days and nights passed by the objectives were achieved one by one. The platoon's roles and taskings were varied: we spent a lot of our time static in overwatch positions; on occasion there would be a short sharp exchange of fire between the insurgents and ourselves. Hours and hours, both day and night, were spent observing and reporting. We also provided flank protection and the usual tasks of route-marking and convoy protection as the convoys brought up supplies of ammunition, water, food and stores to the men and locations that needed them.

Chapter Seven

Endex Operation SOND CHARA

Christmas Day, Lashkar Gah

Operation SOND CHARA was a hard-fought operation for all those involved, even from the opening stages commencing in the early hours of 7 December and continuing relentlessly by day and night until the end of operations on the morning of 25 December, Christmas Day. In particular it was arduous for all the ground troops involved in the fighting, having to live and fight in some extremely unpleasant and ever-changing conditions both tactically and environmentally, being constantly wet, cold and coping with the horrendous mud. This has a physical effect on the body over a prolonged period of time, and just as important as physical wellbeing is the soldier's mind. There is the mental stress and strain involved in fighting at close quarters and being constantly on the go, under pressure, doing things that require a huge amount of effort both mentally and physically to achieve your aim; things that you would not normally do in everyday life, and in an environment of extreme violence against other living beings.

Coping with lack of sleep and food over a period of time has an effect on the body, coupled with the constant uncertainty of your own fate, your own mortality, in the back of your mind, especially when the shit has really hit the fan and you must stand on your own two feet in the middle of it all. There is always that initial shock moment when for a few critical seconds the familiar but horrendous feeling of uncertainty and fear sweeps over your whole body, when sheer violence and chaos reign around you and your comrades in your little world. Then in a flash your mind and body go into response mode, fuelled by instant adrenaline and fear, the do-or-die moment when you must act and respond immediately to a life-threatening situation for yourself and your comrades.

Operation SOND CHARA was a complete success from a military point of view. All the aims and objectives were achieved by all the ISAF forces involved, working together side by side, and the local ANA and ANP made this possible in the area of Nad-e Ali. However, despite the overall success, there were five fatal casualties. The loss of any soldier at any time is wholeheartedly felt throughout our small army, especially by those who served, fought and lived side by side with the fallen comrade. There is a special bond held between soldiers who serve together, a deep trusting friendship that keeps their lost comrade close to their hearts and deep in their thoughts, always remembering them with pride and honour and recalling the good times they shared together. The fallen may not be with us physically but they are not gone: they are with us in our hearts and the sacrifice that they made in serving our beloved little country alongside their comrades, their fellow brothers-in-arms, will never be forgotten.

The whole reconnaissance platoon was still together, even though one or two individuals had received minor bloody flesh wounds. Jimmy (Cpl P) who had received a shrapnel wound from an airburst RPG round to the upper arm and shoulder, and the boss Capt B (MC) who had either a round or ricochet slice open the fleshy part of his cheek just under his eye, were patched up in the field and were still able to soldier on. The only exception was Wild Bill, who had received a single GSW during Operation SOND CHARA and needed to be hospitalized. The entry point of his wound was to the side of his upper chest and the exit point in the shoulder. At this point Bill was back in the UK in a hospital in Birmingham being well and truly looked after and now on the long road to recovery and rehabilitation.

The remainder of the platoon, including all the support vehicles, after regrouping in the early hours of the morning on 25 December rolled out of Patrol Base Argyll en route towards the town of Lashkar Gah in a small convoy, even though one or two of the platoon's vehicles had to be towed or dragged along by the REME support vehicles. Eventually after some hours on the move the platoon finally arrived in convoy without any incidents or breakdowns along our chosen route and made its way into Lashkar Gah camp in the late morning/early afternoon of Christmas Day.

As always the vehicles were parked up or dragged up into call-sign order side by side, close enough to be able to jump from one turret to another.

Tired, weary men dismounted from the vehicles looking in a real sorry state, unshaven and with some individuals supporting rather dodgy-looking '80s' moustaches. All had longish matted hair springing out from underneath their helmets as they removed their skid lids and headsets, exposing some even thicker long sideburns to match the hair on their heads. Most of the men were covered from head to toe in dirt and mud and in some cases torn clothing, mainly to the knee and crotch areas of their trousers. On most individuals from the knees down was a mixture of dark and light-coloured patches of dried mud and staining on their combat trousers, while their boots were still sodden with wet earth and looking very worn. The under-armour shirts were heavily stained with dried whitish sweat rings around the armpit and neck areas and this also appeared down the centre of the back and front of the shirts, running down towards the waist where the shirt was usually tucked in. Many of the platoon had bloodshot eyes, the red veins of the eyeball standing out noticeably against the white backdrop, these tired eyes scanning around as they took in their new surroundings. A thin layer of fine dirt and grit covered their faces and necks and the exposed areas of hands and nails were black with ground-in dirt and carbon from the firing of the weapon systems. Any flesh that had been exposed was brownish in colour and stained, with the grime penetrating into the very pores of the skin.

At this point everyone was unloading the weapons and equipment from the vehicles in relative silence, then some started to allow a little smile to break through on their weary faces on seeing each other and there was some low chatter and muffled laughter that soon broke the tension and cleared the air. After all, it was Christmas Day. That should be enough for the blokes to smile about: hot food at some point, possibly some mail and a dry place to sleep, either on a camp bed or a Thermarest on the floor and under cover away from the elements. After ten minutes or so the guys were back to their normal selves, mucking about and taking the piss out of each other as usual, despite being completely chinned.

The task ahead for the next few hours at least was that of weapon-cleaning, some basic vehicle maintenance and the usual serial number checks and quantity check of specialist equipment under the supervision of Steve, the platoon 2IC, along with the section 2ICs. This routine procedure ensured that all our serial-itemed equipment had been accounted for and checked for 100 per cent serviceability.

A complete de-brief went on concurrently that morning for me and the section commanders conducted by the boss, who then collated all tactical information of importance so he could prepare his final report on the whole involvement of the platoon with Operation SOND CHARA. Then once all the platoon's work had been done and the boss and Steve were happy, it was finally time to get some personal administration done. This started with a chance to get our boots and socks off, get out of our stinking damp clothing and head over to the showers; then have a real shave and finish it all off with a change of clean, dry clothing from our bergans.

Eventually we were all ready. Everything including the weapons had finally been squared away, all personal administration had been completed, and everyone had a bed space in one of the big rubber dome tents over in the transit area to the rear of the camp. We all then headed towards the area of the main cookhouse tent en masse as a platoon. Once we got past the camp CSM who was at the cookhouse entrance just waiting for an excuse to rag the blokes, we all got inside, even Coxy and his big hair and sideburns. We all queued up in single file for our food with Steve, the boss, myself and Pete at the rear; tradition dictates that when it comes to feeding the men, they come first.

We filed through and got a proper cooked meal on a real plate: cold dry turkey, potatoes and vegetables; our first real solid meal in a good few weeks that was not out of a small silver bag and eaten either stone-cold or lukewarm. We all sat down together, momentarily in silence, and a small prayer for the fallen was said before we ate. It was a very late Christmas dinner that year.

Just over a week was spent in this location with the men working continuously on the vehicles, preparing them and the weapon systems for the move back to Camp Bastion. More important was readying ourselves to redeploy straight out on another forthcoming task from Camp Bastion for which the boss had already received his warning order and been given a general outline of the platoon's next tasking.

I myself spent some time in the mornings helping out on the vehicles. The afternoons were given to preparation of my rifle, optics and other equipment with map and available air photo studies of the general area of my next deployment. This included getting information from the operations room and the daily SITREPs and just chatting to members of other units operating out of this camp. This might be details of small or large recent engagements, or small-arms fire contacts with the insurgents, dealing with the obvious first. Were any of the insurgents, if spotted, working in pairs or even in threes and fours? Doctrine in this day and age should be open and flexible depending on the threat against you and the effect you want to cause against a well-trained and motivated enemy in his homeland. Any optics that had been identified, even just binoculars carried on the body, were worth attention or just simply logging in the back of your mind. And how had the insurgents started the engagement? Was it with a well-placed single shot from a concealed fighter hitting a soldier, or an LMG or HMG of some description opening up on friendly forces, or even RPG fire, air or ground-burst, towards a call sign? Possible IED locations such as vulnerable points and choke points were of interest: were they covered by fire immediately or just a few seconds after detonation, or even just before, causing friendly forces to take cover, channelling them to where the IED was located? Recent ambush locations and areas where the insurgents have used come-ons to lead us into a pre-selected killing area; all and any of the above would be of interest. A well-trained marksman or sniper could have a devastating and lethal effect in a very short period of time in any of these situations if employed and supported correctly by his own forces. I tried to get this background information not only from the British call signs on the ground but also our comrades the Estonians and Danish who were operating in the area as well.

All information on the insurgent skills and drills is relevant, down to the level of skill that they employ on a tasking or mission against us, from individual level through to fire team and section level and even larger if anything is seen of the insurgent activity. This would include what they use and how they use it, the ground and the environment, the time of day, the light, wind, weather conditions and ranges; if possible trying to build up a picture or pattern of insurgent marksmanship capability. Obviously the soldier on the ground in the shit at that moment in time and for the remainder of the engagement has a lot more pressing matters on his mind. But once that situation is over and dealt with, people can remember specific things which may be significant and certain points may be more easily recalled by one individual than another.

After ensuring that my rifle and equipment were cleaned and prepared, given a thorough once-over and were fully serviceable, the rifle was retaped and resprayed. I then constructed two new variations of a shroud for the front of my optical scope to conceal its objective lens. Just as important was the updating of my barrel log, shooting record card and personal log book from the short notes that I made out in the field when and where possible. This was basic information regarding the location that I fired from: the range of the engagement, the date and time, weather conditions such as wind strength and direction and the light conditions; almost any information that might be of use in the future was always noted down.

Next was another more enjoyable task that we conducted whenever we could, and that was to get some range time at Bastion. This would commence with a confirmation check zero of our rifles and support weapons to ensure that our rifle's zero was on; in other words that our point of aim onto a given selected target was still our point of impact. We practised elevation and deflection near-miss drills and firing through and around cover. It was important to work as a pair on our communication skills within the pair, such as target identification and prioritization (indexing of targets), firing drills at single and multiple targets and finally practising some co-ordination shooting involving rapid bolt manipulation. Lastly we might get some practice in with 9mm SIG pistols and the 7.62mm GPMG. As always at the end of the range practice there would be some form of small competition between us all with our .338s, usually for the worthy prize of food treats from Camp Bastion NAAFI, all agreed by me, Ben, Matt and Carl before we started.

Chapter Eight

Prelim Moves

Nad-e Ali, Patrol Base Silab

My boss Capt B (MC) had arranged through liaison with 2 PWRR, A Company, who were the theatre reserve battalion and deployed mainly in the area of Nad-e Ali south at the time, that Matt and I would deploy to Patrol Base Silab as a sniper pair and later on be joined by Ben once he had completed another separate tasking with the platoon. We would also conduct a combat estimate from a sniper's point of view regarding the recent increase in insurgent sniper activity or highly-trained marksmen and the threat that they posed, and conduct some form of counter–sniping rather than just try to locate and destroy these threats or minimize their capabilities and effectiveness against us.

We were then to advise friendly forces on how they could deter or reduce this action against them or their location and how they could best minimize the effectiveness of the insurgent sniper/marksman's capabilities from which the threat was now becoming ever more active and effective in this area. This would extend down to the individual level of a single soldier out on the ground to increase his general overall awareness, advising on how to deter the insurgent sniper or marksman from taking him on personally as the selected target, singled out from the remainder of the patrol and then engaged with potentially fatal consequences.

The latter might occur while on a foot or mobile patrol, on sentry duty at a location's main entrance and exit points or on duty in one of the many Sanger locations that are located around every patrol base and are permanently manned. Movement is obviously very restricted within these fortified sentry locations and the individual may be static for some time while providing protection to the base, therefore presenting an easy option to engage for a well-motivated, trained and experienced individual with marksman skills.

The soldier must also consider how active and effective this threat may be during the normal routine and daily running of a patrol base out in a remote and hostile location, bearing in mind that the enemy is always watching by day and night and in all weathers, just waiting for the opportunity to strike. Such tasks as the daily cleaning of the support weapons within their static firing locations – usually on an elevated platform of some description – present a prime opportunity because the men cleaning and maintaining these specialist weapon systems will be the specialists who operate and fire them.

The enemy could be locating, ranging and recording the locations of ammunition and water, food supply stores and the operations room or command bunker structure including satellite dishes, radio antennas or masts and any other form of specialist communications equipment, plus any possible HLS or EHLS locations. The regular movement around these locations and around the compound – for example, distributing stores – and the general pattern of daily life within the base will always mean activity of men and equipment; activity that can never be eliminated. The overall daily patrol and sentry taskings and running of the base must continue, regardless of the threat to the security forces serving there.

The routine changeover of the duty sentries is a non-avoidable predictable pattern of daily routine, just in providing continuous security to the location. So having a covered approach into the Sanger and the use of some form of screen and backdrop in and around its location may be enough to obscure the insurgents' view, or simply just hard-targeting when you think you are in view and getting into the Sanger as fast as you can. Another example is the thunderbox locations which always need attention, the daily clean-up and burning of the faeces being a task for one or two men conducted daily (for health and hygiene reasons). That task is most likely done at the same time each day, again presenting a target pattern at a specific location. And of course, the fact that these locations will be used by all the men and officers makes them an easy opportunist target for a patient, observant marksman. Patrols forming up and waiting to go out or those returning from a tasking provide cracking potential targets to a well-concealed sniper in the surrounding areas of the compound with a good field of view and fire, or one who just wants to put some harassing fire into your location to disrupt the normal routine. Vehicle crews or even just the driver conducting daily checks like the first parading of a vehicle before use and the last after use or conducting any form of maintenance out in the open present another problem as all are usually parked up together in one central location within the base.

Clearly the opportunities to engage such potential targets are absolutely endless within a patrol base or any secure military location. All involve the routine movement of individuals or small groups in a confined location simply conducting their daily tasks, presenting targets of opportunity to an insurgent marksman even before you get out on the ground to patrol and take the fight to him. Just a slight change in the normal daily routine and activities – how and where they are done and prioritizing them in order of importance according to the threat at the time – could be enough to deter the marksman from taking you on that day.

Also worth considering are the hours of darkness and the routines conducted within the patrol base at night including your own light and general battle discipline. The insurgents may not have our calibre of nighttime viewing capabilities or optics for piercing the darkness; therefore they must get that little bit closer for observation during the night hours. Proximity of this kind usually means contact with the insurgents either by accident or deliberately, if you have alert and diligent sentries on duty using thermal image intensifiers to observe into the darkness. This, combined with covering the dead ground using claymore mines and trip flares and making good use of natural or man-made obstacles that the sentry may not be able see into or over, would make it harder for the insurgents to close in around your location. This may then deter them from trying to close down their range and keep them out at further distances where they may be unpractised and less confident to engage.

On occasions, contact with the insurgents at night may be a short, sharp exchange of small-arms fire together with the use of support weapons and some form of para-illumination. In such cases a light plan may be needed to enable the use of other assets such as mortars and artillery if available to support your engagement with HE. What are your counter measures, should this happen? Deter and destroy by day or night are the key words, and even if you never have to use your counter-measure plan, rehearsals of it should always be conducted ensuring that all parties know how to use the equipment safely and effectively. Finally, and very importantly, your CASEVAC plan must be conducted and practised in readiness for the possibility of a man going down anywhere at any time.

Modern technology like improved ground radar, ground sensors, the use of UAVs and other technologies linked up to gun lines or Fast Air and other assets can provide state-of-the-art counter measures to deter or destroy such threats. The training, employment and correct deployment of our own snipers together with lessons learned from previous conflicts and engagements against modern-day insurgents has provided a very well-trained and motivated pool of specialists within ISAF and the overall subject of military combat sniping for these forces.

Most of the expertise within this small area of the military derives from past and recent experience in conflicts and how our snipers are trained from the very beginning. The overall training and development of snipers is set to a very high standard to reach the required level of skill, and is constantly being amended and updated to meet new threats with the use of continuing advances in technology. The majority of NATO forces have truly advanced their sniping skills, moving forward in kit, equipment and training. Also vitally important is the overall combat experience gained on operations, and lessons learned from conflicts in various inhospitable climates and environments are never wasted.

The British, American and Canadian forces have, from my own perspective and experience, always been at the forefront in maintaining the advances in training in all areas of military sniping and all three are very experienced in training future snipers in their respective armies with a rigorous programme providing the skills and fighting experience required for this unique role. Many European forces are now catching up with us in this respect, having very quickly gained considerable combat experience on operations in Iraq and Afghanistan. This progress is facilitated by units conducting cross-training and the exchange of ideas at sniper symposiums held around the world.

Most nations now have state-of-the-art equipment and some truly gifted individuals in this specialized area of the military whose kit and equipment is always being improved upon; these people continually strive to keep themselves physically and mentally at the top of their game. But it is still down to the individual man out on the ground serving in a hostile life-threatening environment, the man in mortal danger, the man with the rifle, to take that shot time and again with no feeling of pity or remorse; just direct lethal action against an incoming threat by the placement of a single well-aimed shot.

The insurgents were starting to increase this type of activity against our locations and patrols out on the ground, engaging both vehicle and foot-mounted call signs, and beginning to make good use of this asset combined with their knowledge of the local area. When the sound of a single shot suddenly snaps through the air, impacting into its intended target and causing a clean instant kill or a messy wounding, that infamous word comes into the minds of the soldiers around the fallen. The word 'sniper' is such a powerful and effective word for a soldier to hear, especially while serving on an operational tour that is both physically and mentally demanding and having to operate daily in a hostile environment against a very proactive enemy. Never knowing when and where you may be in his rifle optics can affect a unit's combat effectiveness and possibly cause low morale due to knowledge of previous encounters and received casualties within the unit. You can seek cover to some degree from artillery, mortars and RPG fire. If you are very lucky and your body armour and helmet fit correctly, you can even dodge a burst from an AK or shrapnel from a grenade. But when a trained sniper or marksman singles you out as his target and places his crosshairs onto you without your knowledge, everything is in his favour. He operates his trigger from a position of concealment at a moment of his choosing when the wind and light conditions are as perfect as possible and the range has been calculated down to millimetres with very little error. The scene is set and so, more or less, is the fatal outcome for the intended target.

Counter-sniping for the hunter, a new sniper moving into and operating on new and unfamiliar ground and in unfamiliar climate conditions must operate with extreme caution and skill, combine all his knowledge and experience and assess his opponent methodically. Wherever possible when gathering intelligence he must not alert the enemy to his own presence, or very soon he himself will become the hunted, a priority target for insurgent snipers. Knowledge of the ground and climate conditions can be gained from foot or mobile patrols and from any location able to oversee the surrounding ground from within or outside the patrol base, either by overt or covert means. It is important to have troops deployed or ready to deploy out on the ground at the same time as the sniper to act as support if needed during the operation. Working as a sniper pair extends from covering each other when conducting taskings to knowing each other well enough to anticipate that one man might do or try something different in order to achieve the aim. Having complete and utter trust in each other and each other's skills is an absolute requirement.

The words and phrases that immediately spring to mind for the sniper to achieve his aim are those such as acute powers of observation, concentration, extreme patience, focus, motivation, adaptability, cunning, deception, lure and a hunting instinct for fellow man. Some targets you may get the opportunity to observe for a short while, all the time knowing at the back of your mind that at any moment you may have to operate that trigger and end a fellow man's life. So your mind, body and will must be fully focused and never hesitant to act decisively at that given point in time. In the deadly game of counter-sniping the end result will be a sniper or sniper pair either living to fight another day or having to pay the ultimate price and being laid to rest. In the words of Ernest Hemingway: 'There is no hunting like the hunting of man, and those who have hunted armed men long enough and like it, never care for anything else thereafter.'

* * *

I had been travelling in the back of a Viking vehicle packed to the hilt with bottled water with all my kit for some hours as part of a CLP en route to Patrol Base Silab. It was a long vehicle convoy: a mix and match of armoured vehicles; heavy-drop vehicles with ISO containers on the back of these monster trucks; the Scimitars manned by our platoon 1 PWRR for the escorts; and the Viking vehicles manned by the Royal Marines and QDG who also provided convoy escorts. There were also Vixen Land Rover types (very similar to the old Snatch vehicles), packed full with men, kit and equipment. Matt my No 2 was travelling in one of these and providing one of the top cover sentries. These re-supply convoys always carried ammunition, water, rations and anything else that the patrol bases needed or asked for, if the RQMS and CQMS could get their hands on it. It was all loaded onto the convoy for the troops, together with any mail from Camp Bastion for the outstations. These convoys were a lifeline for the patrol bases, carrying a huge amount of kit, equipment and manpower if needed to one location in one long CLP and usually in all weathers.

The use of helicopters was limited in the role of re-supply due to the amount and type of stores and equipment that could be carried by the airframe as DAC; items classified as Dangerous Air Cargo that could be a potential hazard to the crew and airframe. A few CLP convoys could undertake this role and free up the airframes for other taskings. But the main factor was that re-supply was not the helicopter's primary role and the ever-changing weather conditions in the winter months could rapidly change for the worse, making flying very dangerous. That could make a big difference to a planned re-supply operation being able to fly or not. However, if there was a casualty on the ground needing an urgent CASEVAC or if close air support was required somewhere the pilots and crews would be in their machines, up and flying and coming to your assistance no matter what.

Patrol Base Silab as it was known when we turned up was previously known as Patrol Base Barbarian. PB Silab was one of a number of bases that were spread out over the Nad-e Ali area and some of them were built in these specific locations as a direct result of operational success, i.e. the ground that ISAF forces had taken back from the insurgents during Operation SOND CHARA. This base was manned by soldiers from our sister battalion, 2nd Battalion The Princess of Wales's Royal Regiment, A Company, 3 Platoon who were the theatre reserve battalion based in Cyprus at the time.

The convoy rolled slowly into the patrol base one by one and behind each vehicle as it moved over the fine-grained surface followed a fine cloud of desert dirt and sand, engulfing the convoy and the soldiers providing top cover at every turn or manoeuvre of the vehicles snaking their way towards the intended destination. Finally the vehicles came to an abrupt halt one by one within the security of the compound walls of the patrol base in a dust cloud, further engulfing them and the top cover sentries. The vehicles were parked in the order of march so they would be ready for the move out once the stores and equipment for the patrol base had been unloaded.

The main gate was closed and the ANP went back to manning a VCP outside the front gate on the mud track. The vehicle crews and their passengers dismounted, removing their body armour, helmets and headsets and more or less immediately either went to relieve themselves or throw some water down their necks. However, some unfortunate crews had to crack straight on with vehicle maintenance once they identified what was wrong and start the repair work in order to be ready to move out with the rest of the convoy once it was unloaded and move on to the next location or head back towards Camp Bastion. Commanders had their mapping out and were standing huddled together in small groups discussing the next part of the move; the route and actions on for the next phase. People were cutting about everywhere all over the base and the unloading started more or less straight away of everything required for this location before the convoy could move off to its next destination.

I got my kit out of the back of the Viking vehicle in which I had been travelling, stacked my bergan, day sack and man bag next to a wall and grabbed both my rifles, the A2 and .338, which I slung over my back and thanked the crew for getting me here, then went off to find Matt somewhere along in the convoy. I walked across a large open area of ground within the patrol base. The whole base was completely surrounded by a high thick wall, in parts made by the previous local occupants and then the gaps filled in with some large sections of HESCO to completely seal off this position from the locals. Built up by the Royal Engineers, prominent purpose-built HESCO-fortified Sanger positions rose up out of the ground, providing slightly increased elevation over the height of the main compound walls to oversee the ground, the general area and compound buildings located nearby and surrounding the patrol base.

There were a few small single-level roofed outbuildings that must have been built a very long time ago when they were occupied by local farmers and these were located roughly in each corner of the main compound and at central locations along the long sections of wall facing into the village which I could see from the centre of the patrol base. I could also see through some of the large holes and collapsed sections of the inner walls of the compound that they were connected and surrounded some of the small outbuildings in which the troops lived and used as platoon stores.

One of these small outbuildings would soon become my and Matt's home for a while. It was located over in the MSG corner where there was also a well-constructed and elevated Sanger which had a GPMG set up in the SF role and a Javelin missile system set up ready to go as I found out within half an hour of arriving. We had soon been shown around and given a ground situation and ground brief by our new boss Lt D who was the rifle platoon commander, the OC for 3 Platoon. He himself was on his second deployment from Cyprus to Afghanistan as part of the theatre reserve battalion.

I found Matt and walked over to his vehicle, ragging him jokingly about how the hell could he be used as a top cover sentry and be able to see out over the top cover hatch as he got out of the rear of the vehicle covered in dust and grit. He then got even more of a ragging when I went round to the rear of the vehicle to help him with his kit and equipment, as I could see that in the back he had been standing on top of the spare ammunition containers just to be able to see out of the top cover hatch. As always he was like a lightning bolt with a reply in the same manner, which always made me laugh to myself. We got his kit together and went and dumped it next to mine against the wall.

People were moving around everywhere in this location. The platoon that were resident in the base were quickly cutting about under the direction and watchful eye of Sgt M, Moggy and his JNCOs, helping him to supervise the men as they unloaded the stores. Moggy was the rifle platoon sergeant of this location; he was also an experienced sniper instructor who also just happened to have his L96 with him.

We were approached by Cpl Spenny P who was from machine guns and he introduced himself to us. He had been deployed to Afghanistan before as well and was a veteran from Roshan Tower where a small force from A Company 2 PWRR were under continuous attack both by day and night for nine days from small-arms and RPG fire and relentless mortar fire from a vastly superior number of insurgents but had managed to hold the ground and the small position on top of the hill throughout the siege of this location. We grabbed all our kit and equipment and were quickly shown to where we were going to be living and working. We were to be co-located just behind what was known as the MSG Sanger in which the Javelin missile system and GPMG in the SF role were set up and ready to fire and there was enough room to be able to squeeze me and Matt in to fire from this position. We also had the option of gaining several more metres in height by moving out and up onto the roof of the Sanger itself.

The roof had a knee-high sandbag wall going around it that could offer us some protection from small-arms fire and gave us a larger field of view, opening up our arcs of fire and observation to almost 360° around us. To the left of our Sanger location we had four 105mm light field guns that were manned by the Royal Artillery and a few Australian gunners who provided fire support, either HE or light and smoke, when needed in support of troops in contact out on the ground. They also supported the patrol base if they could when this location came under insurgent attack.

We were taken through an entrance with a handmade sign above it with a freehand drawing depicting that Jav Cav (my name for Big Sam, short for Javelin Cavalry who could always be relied upon to save the day if it was going wrong with a missile or two stopping the insurgents) and snipers lived there. This opened up into a small courtyard, on one side of which was a small compound building occupied by Spenny and Joseph, who was also SF-trained and experienced, and they had made it as comfortable as they could. They maintained and operated the SF gun, and Big Sam who was from Javelin platoon and was the operator of that system lived there with them. In a corner on the other side of the small courtyard was an even smaller outbuilding where Matt and I would make our digs for at least the next few weeks.

Some of the environmental factors were slightly different from our last location and its surrounding areas. The wind was stronger and more unpredictable especially during the late afternoon period, picking up in strength from about 1500hrs and continuing through until just after last light. We started the routine of many hours of observation, noting down any points of relevance to do with the local pattern of life, the ground, areas of natural foliage, man-made structures and natural obstacles within our arcs of observation and fire for future reference.

We had to try to establish some sort of a pattern for the wind during the late afternoons out at the ranges where we could be expected to engage the insurgents and we also had an added factor in that the season was slowly starting to change and so were the natural light conditions but the main concern was the much more unpredictable wind. However, the light conditions were improving and so was the overall air and ground temperature. This, especially out at the middle and far ranges, produced mirage which when present was a good aid in judging the wind speed when viewing through our optics. We could sometimes assess the wind simply by observing the natural foliage growing in the local area and the black flags on top of the graveyard – basically observing anything that was natural or man-made that might be wind-affected – and try to use this as an additional aid.

We must consider the wind at the firer's end, out in the middle distances of the overall shoot and at the target end. The very slight changes in wind direction and strength and other possible factors that are outside of the firer's control such as humidity, temperature and rain can have a minute effect on the round from the point of operating the trigger to the round then leaving the end of the rifle barrel on its short journey to the target end, and of course may affect the round as it travels through the air.

Firing from inside an urban environment or into an urban setting from a position on the outskirts of a town or village is another vitally important factor for the sniper to consider. It is not like firing out on a range or field-firing across open ground. The wind is affected very differently in an urban setting where it may become channelled by the man-made structures and natural features inside the village or town and then increase in speed, becoming stronger or gustier within the spaces between solid structures and therefore much harder for the sniper to estimate.

Also factored into the sniper's calculations before taking his shot onto the target is his final firing position, whether it is elevated or not, and if so understanding the applications of fire regarding high-angle firing and the use of the Angle Cosine Indicator which is attached to the side of the weapon and used in conjunction with the relevant data cards. The Mildot Master (a protractor-shaped device with a scale and weighted string or strong cotton used to measure the angle of the rifle barrel) is another implement used in conjunction with the Cosine Indicator in determining the angle of the shot.

Just a few more facts and figures to be added to the overall firing calculation; data that if correctly calculated should put a single well-aimed round down towards the target end and impact the target with the desired effect.

Chapter Nine

Nad-e Ali, Patrol Base Silab

Part I

I awoke suddenly, bolt upright in my sleeping bag in an instant, instinctively swung my legs over the side of my cot bed as I was now sitting up and tried to free myself from the damn thing in complete darkness. I was startled and shocked by the alarming, piercing sounds of erratic small-arms and HMG fire that had started just moments before and the intermittent sound of shouting voices coming from outside the small outbuilding in which Matt and I had now made our digs. Next the sound of the para–illumination rocket being fired filled my ears with its familiar whooshing sound as it travelled almost vertically up into the darkness of the night, leaving a whitish smoke trail in its wake lingering in the sky on its line of trajectory. The rocket then made a dull popping sound, immediately producing a brilliant bright white light as it was carried along high in the night sky by the wind and floated along, burning merrily away. The ground below was vividly illuminated for a few moments, dancing shadows appearing everywhere as the light passed by, followed by the return of darkness. One or two further rockets were then fired to illuminate the general area out to the front of the MSG Sanger location.

It is a very uncomfortable feeling to be woken like that, thrown straight into a hyper-alert state from the depths of a comfortable, calm sleep. The sudden and instinctive rush to go for either your rifle, helmet or even body armour, trying to get it on or at least at the ready, is followed by an instant whole body sweat, a dry mouth and a feeling of nervousness that sweeps all over you. In that very short time since waking, your heart has gone into super-pounding mode while you are thinking: 'What the fuck is going on? What and where are they attacking? How many of them are there? Have they managed to get inside the perimeter?' The mind races through all possible scenarios while gripped by the feeling of being caught out, off your guard and vulnerable for several moments.

I looked over to my right to find Matt awake and going through the same response: we were both trying to get our kit on as quickly as physically possible and once we had a little light from our head torches it was easier, especially when trying frantically to lace up our boots. Once we were both ready and good to go we grabbed our individual A2 rifles and nighttime viewing devices and made our way straight to the MSG Sanger. On our short route we could see overhead the red and green tracer that was flying over and around the location, zipping through the darkness making snapping sounds as the individual rounds travelled through the cold air and then made contact with a solid object, impacting into the surrounding compound walls or outbuildings.

We both made it to the back of the MSG Sanger and scrambled up a mixture of HESCO stepping and wooden steps into the rear of the Sanger. Joseph was in the standing position in the far corner and in control of the gun. He was letting rip with the GPMG and an almost continuous red line of 7.62mm in short bursts flew towards the insurgent firing positions that were firing furiously back. Tracer rounds were flying everywhere: the two-way range was certainly open that night. Suddenly we heard a familiar whooshing sound and observed the small signature flash of an RPG as it was fired in the darkness. In that split second, from slightly right of our location – which looked very fucking close in range to us in the Sanger – the warhead round was zipping its way towards us as it flew through the air, then slammed into the lower outer wall and detonated with a blinding flash, sending lumps of dried wall, debris and dust showering down on our location.

Spenny was kneeling down and handling ammunition containers, opening another box of 7.62mm link with his multi-tool and frantically pulling and ripping out the packaging to reveal the belted ammunition. He started to prepare the ammunition for Joseph to fire, all the while trying to back-brief us on what the fuck was happening, and Big Sam was observing, scanning the ground and forward compounds out to our front through the CLU.

Matt and I were in the standing position observing through our night-viewing devices scanning over the ground out on our flanks and we both could see that Sanger Three was being engaged and had also opened up, returning fire back towards the insurgents and letting rip with their GPMG and rifles out to our left flank. I was mesmerized for a few short moments by the muzzle flashes from the firing and watched the tracer rounds being spat out from over in that dark corner of the patrol base going towards the insurgent firing positions. Small flickers of light illuminated the darkness as the Sanger sentries fired their weapons: a mixture of single-shot 5.56mm and 7.62mm burst fire from the other light role GPMG. We both then started to make use of our SA80 A2s and could discern some muzzle flashes in the darkness out to our front coming from the direction of the forward edge of the village. Observing through our nighttime viewing optics, the target area was green in colour and we could clearly make out the ground and its contents with a little fine focus here and there; they were spot on.

Joseph and Spenny worked the GPMG as a pair; there were empty cases and link beginning to go all over the Sanger, and Sam was telling us what he was observing as he used the CLU to observe into the darkness and try to pick up heat signatures. The fire-fight continued for about a further twenty-five minutes: it would suddenly die down and give both sides a brief respite to sort themselves out and then erupt again just as it had started with an RPG warhead or two zipping its way towards us and a long burst of automatic fire possibly from an insurgent LMG, then what must have been the remainder of the insurgents opening up and joining in. Each time, however, they had moved positions and regrouped to continue their attack from a new direction but it didn't take long to identify this and deal with the threat.

When involved in any form of contact, for me time is irrelevant: it goes by so quickly regardless of being tired or in some form of physical discomfort; you are constantly thinking and moving, just doing what you have to do to get yourself through it. All your training and experience comes into play at some point; also making maximum use of whatever resources are available to you at the time to eliminate the threat. Adaptability and flexibility are the best aids in your thought process in estimating a given situation and how you are going to deal with it.

The great thing is working as a member of a team, for the team, with your battle buddies all having the same end goal in mind, knowing that at some point it will end and hopefully we will all get to sit back at base in a small huddle around the Hexi blocks and mess tins or cooker, heating our boil-in-the-bags and making brews, talking shit about the day and the contact.

We'd be ticking off another day closer to our end of tour date or closer to our R&R; maybe the countdown to returning to Camp Bastion and starting the handover of all the kit and equipment to the incoming unit, then just waiting to get on the flight out of theatre on the big bird and home.

The night attack by the insurgents came to a gradual end but only after some more heavy exchanges of fire between us in which we clearly owned the night as every man had some form of night-viewing device to use and this gave us a real advantage over the insurgents in being able to locate and effectively engage them. Virtual silence then returned, broken only by the howling of the wild dogs, and darkness took control of the night. We remained in our position alert and in total silence, with a quick magazine change and a fresh belt of ammunition loaded onto the GPMG ready to go again if needed.

Everyone remained in watch-and-shoot mode, observing and listening, ready and waiting should the insurgents strike again. But nothing. Sam was not picking up anything to do with human movement while observing through the CLU, only some wild dogs that were roaming around the area scavenging for food. So we eventually stood down, cleared up the brass and link from where we had been firing and cleaned up the Sanger floor from all the waste packaging from the H83 containers. At last the sentries were left to continue with their duty in silence, maintaining the security of the patrol base and settling back down to the nighttime rota of rest and duty.

The next few days passed by with many hours of observation and familiarizing ourselves with the local pattern of life around our new location, also gathering information on the ground while out on foot patrol. We settled down into a steady routine in the base with a simple personal administration routine within our small group. Matt was the head chef who cooked the food and made the evening brews daily along with Spenny and Joseph, knocking up some cracking all-in-one meals for us. The guys had dug a small deep hole and made a cooking area using some HESCO wire as an improvised grill with Hexi blocks piled up nearby ready to use, and they filled a few sandbags on which we could sit around the cooking hole. It was my job to do our laundry. I got a large black plastic end cap of a transportation tube for the Javelin missile that served as a good wash bowl and I made a HESCO wire stand on which I could hang the clothes out to dry during the daytime.

All our spare ammunition, ammunition for our sniper rifles, water and rations were collected and stacked up inside our small hut by the entrance and then our small MSG grouping set to reinforcing a few sections of the HESCO wall and main compound wall that were damaged and needed to be repaired. This was a task that could possibly take several days to complete and we also had to sandbag the roof of our small outbuilding. The Royal Engineers came to our aid in this task and also put up wooden support beams inside to strengthen the roof in preparation for the additional weight that it would have to bear. The 40mm GMG was going to be relocated up there to be able to cover more vital ground leading out to and beyond the compounds from where the insurgents were frequently engaging us, thereby improving our overall defences.

An oil that I had been issued with my brand-new .338 worked wonders on the gas parts and working parts as we soon found out on the GPMG, removing carbon from excessive firing of the gun with ease and it was soon shared around the other guns and the GMG in our location to help with the cleaning of these weapon systems. I had also been issued with some new lens-cleaning fluid and lens brushes that helped me to maintain my optics. The wind was now getting stronger in the afternoons and the fine grit and sand got everywhere, seeming to collect very easily on my rifle and particularly in the exposed areas of the elevation and deflection drums, around the focusing ring and anywhere on the scope and rifle body that was not covered. This needed to be brushed away regularly when it built up to allow the free smooth working of both the rifle and the optical scope.

We had been preparing our kit and equipment and collecting data that would help us in our forthcoming taskings with the platoon and deployment out on the ground as an MSG and from recent intelligence, information had been gained that there was an increasing and active threat from insurgent snipers in our local area of operations. For me there was a final bit of detail that needed to be attended to. I needed to replace my weathered cut-down sandbag used as a helmet cover that was placed over my normal issued helmet cover but which I could easily remove if necessary for friendly forces identification purposes. For me, the use of a weathered sandbag was ideal: the colour and texture matched the elevated sandbagged positions from which I often observed and fired, so I wanted to blend in as part of the sandbagged feature. I had briefed men going out on local foot patrols regarding my exact location so they could look back towards me from the near, middle and far distances and see whether I was visible. I had used this effect before on the rooftop position at PB Argyll and wanted to know how easy it was to be observed using the natural eye and issued optics. The idea seemed to work as desired.

For our final confirmation Matt and I would go out individually as members of a patrol and conduct our own observations from different ranges and angles or fields of view to clarify what the insurgents might be able to see of us. We would also identify on the ground their potential firing positions: areas on the ditches in and around the compounds with clear lines of fire and sight, even out on the slightly elevated, undulating ground that surrounded our PB at the further ranges. It was also necessary to know what could be seen of the sentries inside the Sangers and whether we needed to put a drop inside the Sanger to create a backdrop, obscuring the silhouette of the standing sentry while on duty during the daytime period, especially with the now improving light conditions.

My final idea for a basic and effective aid to concealment was a simple one and readily available around the patrol base. I found a suitable-sized piece of material that could cover my rifle, head and shoulder area and the trunk of my body if necessary; an item that could also fold down small enough to fit inside my day sack and was lightweight enough to carry even when covered with dry earth and grit. A weathered piece of the cloth that lined the inside of the HESCO wire baskets cut out in an irregular shape large enough for the job was perfect for my intended use. Just as Russian and German snipers during the Second World War improvised and used their own country's respective plain or camouflage pattern versions of the now familiar modern-day poncho as an aid to concealment on sniping tasks in the ruined city of Stalingrad, I would use my thin improvised cloth poncho in the same manner.

It did not take long to confirm the fact that an insurgent sniper was now operating in our area. The platoon had been out on a number of joint foot patrols with the ANA and each time the platoon was involved in a number of small-arms contacts varying in the means of initiation and the use of insurgent firepower; the insurgents had started to combine the use of airburst RPG with LMG fire. A few days passed by uneventfully, and then the platoon and a small detachment of ANA went out on another routine joint foot patrol. After half an hour or so of being out on the ground, the interpreter sure enough picked up the voices of the insurgent commanders coming over the ICOM scanner and passed the information on to the boss, Lt D, on what was being said. He alerted him to the fact that the insurgents were now already in position and ready to attack. They could observe the patrol and were just waiting for the order from their command structure over the radio to initiate the attack.

The platoon went firm with the men finding cover where they could, taking a knee and using their rifle optics to observe the ground around them of their individual arcs and trying to PID any insurgent or suspicious movement around the area. But the local farmers had disappeared from working in their fields and off the narrow mud tracks including the main track that passed through this area. It too was clear of movement: the activity was minimal from the locals who would usually be herding their animals along or transiting these routes at this time of day.

The next few minutes seemed to pass by slowly with everyone on edge, just waiting and anticipating what was about to happen. Then out of nowhere the sound of small-arms fire filled the air. The platoon was in contact again and suddenly the front of the call sign was pinned down by a heavy rate of small-arms fire from both single-shot and burst as it came zipping in towards them. An initial RPG was also fired towards the front of the call sign coming in high from somewhere over in the middle distance and exploding in mid-air just short of the lead men, sending fragmentation out everywhere up at the front. The incoming fire was from a number of single-storey walled compounds and irrigation ditches that now had plenty of thick vegetation growing in and around them. This made an excellent natural screen that the insurgents were using to fire from and manoeuvre around; its depth, trapped shadow and backdrop made target identification much harder with very limited muzzle flash and smoke signature at the insurgent end and above all offered them protection from our return fire.

The platoon quickly organized itself and set about extracting from the killing area and laid down a heavy rate of return fire combined with a heavy smokescreen to mask its extraction from the present location. Then the men started to move and fight their way back towards the direction and security of the patrol base. Firing and manoeuvring all the way, up and down in short and sharp bounds, the men were having to move over wet mud and through thick foliage, their kit and equipment weighing them down as they struggled through the mud and crossed small irrigation ditches filled with water. So much effort was needed each time, just to get moving.

The antennas on the ECM equipment that was carried by some individuals could be seen whipping around violently in the air as those men moved from one piece of cover towards the next. Those antennas could also be seen sticking out of the top of the long grass that grew in and around the ditches as the men tried to cross these obstacles. Each and every man was constantly looking around and shouting out to one another: where each man was on each bound, ensuring that he had moved and got the message, and the young commanders were working their arses off trying to maintain command and control at their level.

The boss co-ordinated all his available firepower, using everything within the platoon out on the ground and via the radio as he called for assistance from other assets that were available to him while on the ground for that task. He was supported from within the patrol base by the 105mm light field guns manned by the Royal Artillery and the GPMGs in the rear two Sangers that could observe over the ground towards the area of the compounds and irrigation ditches to help with the platoon's extraction. The 105mm guns were soon in action and firing HE down towards the area of the compounds. The sound of the guns firing and the high-explosive shells impacting and detonating at the target end was awesome to hear and watch as brown clouds of debris were thrown high into the afternoon sky and then came falling back down to earth in a few short moments. The dust cloud that had been created by this effect just hung in the sky, lingering briefly until dispersed by the wind.

Every single soldier, every man that day dug out blind in professionalism and individual courage while under fire during the time in contact. However, the platoon did receive a casualty. A member of the ANA had received a single gunshot wound to the head and collapsed to the ground immediately where he had once stood among his ANA comrades during the fire-fight. During the critical minutes that followed it was certain that he had received a fatal GSW to the head and that death would have been instant: the damage caused to the head was catastrophic. As the situation developed in those few vital minutes and the incoming small-arms fire seemed to intensify all around them, there was only one option. The platoon had to extract our fallen comrade off the ground and back to the PB and Joseph was tasked to bring him back. Joseph is a physically strong man and with courage to match his physical and mental strength. The whole time while trying to manoeuvre himself and the fallen comrade he was still under contact from the insurgents, wading through mud and water that was chest-deep in places while moving along the irrigation ditch that was used for part of the platoon's extraction route. The ditch led back towards the direction of the patrol base and offered some degree of cover from fire and view on the return journey. Meanwhile throughout the contact the two rear Sanger locations continued the engagement against the insurgents and the 40mm GMG joined in to cover the extraction.

Once everyone was in through the front gate the ANA quickly took their fallen comrade away to the area of small outbuildings which they occupied in the base, clearly visibly shaken by the day's events. A quick head check was conducted and the section commanders accounted for their men and reported to Moggy. The men looked totally physically and mentally fucked: hot, sweaty, wet and covered in mud. A SITREP was prepared by the boss and sent as soon as all the information had been collated to higher formation regarding the events that took place that day.

Joseph told me later that evening about his experiences during the day and described the events in considerable detail as we sat in the darkness with just a small amount of light created by the dying flickering flames from a few Hexi blocks that were left to burn out after we had made a brew. He said the body was still warm, with a mixture of blood and brain matter slowly and continuously oozing out of the large exit wound to the back of the upper skull. The lifeless eyes were still open and looked as if they had clouded or glazed over and the lower jaw was stuck in a half open manner, rigid with the bottom lip receding to reveal the lower teeth and gums. The remainder of the body was limp but felt heavy like a weighted rag doll and after some time in the cold water the weight seemed to increase. We spent a good few hours that night talking about what had happened during the contact and the fact that it was not just a question of a lucky random hit. After the platoon had extracted and everyone had been back in the patrol base for about an hour, the interpreter told me that the chatter coming over the ICOM scanner was the local insurgent commander congratulating the sniper who had taken the shot and got a confirmed kill.

I deduced that it must have been a 7.62mm-calibre round due to the exit wound size and the physical damage caused to the skull, very similar to the 7.62mm round fired from the L96 British sniper rifle. The rifle most readily available to the insurgents was the SVD Dragunov 7.62mm sniper. It was likely a first-round hit, penetrating cleanly straight through the top of the skull with a clean exit wound to the rear of the head. From the information Joseph gave me about the location of the possible firing or contact point to the exact location on the ground where our comrade had fallen, it seemed that the range was close; possibly only about 300 metres or slightly less.

It was also possible that this insurgent sniper or marksman worked within a small combat grouping of sorts for added protection and while all his insurgent comrades were engaging us, used all the surrounding noise and activity as a diversion or screen to mask his own actions. This would have also allowed him more time to prepare, select a firing position and just as important, spend time in observation of us and the contact, observing our reactions to the current activity and possibly prioritizing his target selection so that when he came to finally fire that single well-aimed shot, it was a fatal one. Time was on his side as at this point the insurgents had the upper hand. All the surrounding activity would have afforded him a little more freedom of movement to obtain a better field of view and line of fire onto a target of his choosing.

The next week or two seemed to pass by slowly. The days were growing gradually brighter and longer as the temperature started to rise and the routine foot patrols both by day and night would often end up in some sort of fire-fight against the insurgents. The MSG grouping would be in a position on the ground that was able to provide overwatch and fire if needed to support the patrol should it come under contact. We were the eyes and ears of the patrol: anything of relevance was immediately passed on over the radio to the boss and to Moggy.

Today's patrol had been a quiet one: even the civilian activity was unusually minimal and there was no contact with the insurgents during the whole patrol which might have been due to the day's poor weather conditions that were quite bad compared to previous days. The overall visibility of the ground and compounds around our location had been greatly reduced due to the wind suddenly gaining in strength and picking up dirt and sand as it travelled across the ground, carrying it along in the air like a mini sandstorm. The cloud cover was low, which also reduced the amount of natural sunlight.

We had been back in the patrol base for some hours, staying inside the protection of our small hut and out of the bad weather. The afternoon passed by slowly, with Matt and I just reading or listening to our iPods while resting on our cot beds. Then, according to my firing log book, at about 1700hrs all hell broke loose again with the PB and in particular the MSG Sanger coming under attack from a mixture of small-arms and long burst fire, most probably from an LMG. Immediately both of us reacted by springing up from our beds, grabbing our body armour and helmets and getting them on as fast as possible.

We grabbed our rifles and optics and Matt seized the laser range-finder, although the weather conditions today were far from ideal for using this piece of equipment. We ran our now familiar route to the MSG Sanger in daylight, past the narrow gap between the side of Spenny's much larger and more comfortable pad and the main high outer protective wall of the patrol base. Joseph and Spenny were just coming out of the rear entrance to their accommodation with all their gear on and heading towards the steps into the rear of the Sanger. The desert camouflage netting that we had put up around the Sanger location and several metres of the approach route into the rear was flapping, straining under the strength of the wind. This was held in place by 6ft pickets and even the usual sagging sections of netting were taut in the wind.

The insurgent fire was coming from several of the usual firing positions, varying in distance out to a maximum range of roughly 500 metres from our position inside the Sanger. It was issuing from the inside of windows and doorways, from the rooftops of the compounds and from ground level around the irrigation ditches and naturally foliaged areas by a small group of single-storey compound buildings between 370 and 420 metres away. These were surrounded by a high solid wall over 10 feet high in sections with pre-prepared firing holes ('murder holes') at various points along its length from which the insurgents had already regularly engaged us. We had previously ranged these locations and frequently engaged the positions so the range was not a drama; we just had to confirm our wind data once a target had been identified.

However, the strong and unpredictable wind was not as it would usually be around this time of day. These unusual conditions plus observing in such poor light meant that a little work was required for positive identification of a target and collation of data to be able to engage effectively. But for certain I would be firing from the standing bipod-supported position which was comfortable and sustainable and allowed some freedom of movement with clear fields of view and fire.

The Sanger was full of activity. There were now five of us in this location and it was a little tight. Target indications were being called out over the noise of the GPMG and rifle fire with the GPMG firing rapid burst fire, spitting out a continuous arching line of red tracer and ball from the end of the smoking flash eliminator of the gun and ejecting a mixture of link and hot empty cases. Matt and I were squeezed in a corner of the Sanger, more or less shoulder to shoulder. I was in a sustainable position to be able to observe through my rifle scope and fire, with Matt to my left. He was using his binoculars, frantically scanning the area of the compounds out to our front and switched to the spotting scope on seeing something of interest for confirmation.

I was moving around slightly in my position, scanning and then abruptly stopping to observe something of interest that had caught my eye. Due to the movement and colour I thought it to be a muzzle flash just glimpsed through my optics. I quickly adjusted my sight picture by turning up the magnification ring from x6 until what I was trying to observe became clear enough for me to identify and then made fine adjustments to the focus ring and parallax adjuster to get the clearest possible picture. But whatever it was had now gone and I found myself frantically scanning the dark foliage for a few more minutes, hoping to spot some further movement.

Fire from the insurgents was still coming in towards our location and impacting into the front-facing wall of the Sanger and the open ground just below us out to our front. Rounds were striking into the dusty ground making a snapping sound on impact and some found their mark in the overhead protection, the roof area just above our heads. When you hear that you instinctively just bow your head or get low for a split second, thinking: 'Holy shit, that was close.'

Moments later Stu who manned the 60mm mortar had been given a fire mission and started to drop HE mortar bombs onto the area of the compounds from where the fire was coming. As each round detonated on impact, its dull crump filled the air and echoed over the small-arms fire. Clouds of dust and debris were thrown up from the ground behind the wall, whisked up high into the late afternoon sky and quickly dispersed; after observing this for a few minutes we were better able to work out the wind's strength and direction. Our thoughts were that it was a strong, gusting wind basically going from left to right and when not gusting, it was a light oblique wind which should not be a problem at this range. But there was no time to fuck about, especially not at this range, trying to decide if we should add the corrections for a little error of an oblique wind: we just needed to PID something and then get some accurate fire going down towards the target end.

The insurgent small-arms fire was still coming in towards us at a good sustainable rate and observation of them was proving hard; only on occasions were we able to catch a glimmer of movement or the yellowish and orange mix of colour from their muzzle flash. Sometimes these flashes were noticeable only for a few seconds, and by the time you had reacted and aligned yourself and your rifle in that direction and were in a position ready to fire, they had moved and were gone.

The detonations of the mortar rounds that were now raining down and impacting over on the other side of the wall surrounding the compounds must have made the insurgents a little more jumpy and edgy. More movement was observed and now the telltale signature muzzle flashes were more prominent and more easily seen from their firing positions within the cover of trapped shadow provided by the doorways, windows and holes in the walls. Matt's eyes and mine were staring through our optics, straining to see into these dark locations as well as scanning the forward edge of the foliage screen to our front. The irrigation ditch was always a favoured insurgent firing position, even though it usually got riddled with HE from the pounding of our GPMGs and the 40mm GMG.

I had already put wind and range corrections onto my rifle scope which I set at the range I had estimated for some of the insurgent firing points: this was about 420 metres away at the closest point of contact from our Sanger location. Observing over into this familiar field of view, these possible firing positions were located in and around the areas close to the compounds and ditches that were engaging us and were well-used and known insurgent contact points. I had already set my deflection drum and thought that even if I was just slightly out, the error at this range would be very minimal: just over 40 millimetres or so. I reckoned I could cope with that and make any minor adjustments if I really needed to after firing the first round, observing the point of impact, making the necessary corrections and then with a bit of rapid bolt manipulation, quickly re-engaging the target.

The whole area was now being hammered by small-arms fire and the HE rounds from the 60mm were having an effect on the ground around the compounds. The operations room had been in comms with Camp Bastion and an AH [attack helicopter] call sign that was in the local area was diverted: once again the Apache helicopters were en route to our location for additional fire support. The insurgents were still putting down some fairly accurate fire at a rapid rate towards us and I was drawn back into observing a foliaged section of the irrigation ditch slightly to the left of a prominent gap between one compound wall and the side of an adjacent small compound by the sight of short bursts of muzzle flash coming from there.

I brought Matt on to what I was trying to observe and he switched to using the spotting scope for greater detail. I went back to observing the ground area in and around the forward ditches, and after some frantic minutes of observing and scanning, there it was again: that signature muzzle flash in roughly the same area. However, this time the shooter had moved ever so slightly forward from his previous position and exposed himself just a fraction more. That was all I needed.

I could now make out the head and shoulders of the insurgent as he was firing in the prone position using the irrigation ditch for cover from view and fire, or so he must have thought. I aimed straight away onto the centre of observed mass, as was the usual drill when this much of the target is visible.

I simultaneously removed my safety catch and held that point of aim for a few short moments, confirming with Matt what I was observing and making a final adjustment to my fine focus. Then I slowly and deliberately operated my trigger, squeezing it gently, pulling it to the rear with my right forefinger and fired a single ball 8.59mm round.

The round impacted into the target just below the shoulder area, the very top left part of the upper chest, and the force of this impact appeared to penetrate into the insurgent's body with some power, forcing his arm out and upwards to the rear, simultaneously turning him briefly onto his back. The target then seemed to slide down and disappear out of sight and must have slipped to the bottom of the irrigation ditch, his rifle still lying on the forward edge of the embankment at an angle with the muzzle pointing upwards.

I chambered another round, maintaining my sight picture of the target area and taking in long, slow, deliberate breaths just to remain calm, keep my pounding heart at a regular rhythm and concentrate on what I was doing. After a few more minutes of observation through my rifle scope I could not pick out or ping any sign of movement in the area. I had to release my right hand from its hold on my rifle and break my position as my hand was starting to go numb and I had pins and needles in my right arm. I rested the rifle butt on the parapet of the Sanger and shook my arm, trying to free myself of these unwelcome sensations.

We could now hear and see the Apache helicopters approaching, a pair coming in from over in the far distance. The 60mm mortar had now stopped firing and there was just small–arms fire going out and coming back into our location. The Apaches were like a pair of hunting eagles working together, prowling the skies above us and then suddenly striking, unleashing their ordnance and ripping up the ground in and around the area of the compounds with devastating effect. We simply watched in awe for a mad few minutes the firepower being unleashed on the insurgents in this small area. After a short exchange of heavy fire everything seemed to gradually calm down, becoming more sporadic and less frequent until it stopped completely on both sides.

Near silence then followed as the dust and debris settled back down to earth where nothing stirred at all apart from a few wild dogs that came running out from behind the cover of the pounded compounds and kept running back towards the outskirts of the village. All that could be heard was the engines of the attack helicopters as they circled or hovered above us, waiting and watching and ready to strike again, but still nothing stirred in that area.

It was a cracking job done and one hell of a thing to witness the Apaches in action right in front of us. From Matt's and my point of view they had helped open up our arcs of view and fire slightly in a certain direction by destroying some of the low-lying man-made and natural obstacles that obstructed our field of view. It worked both ways though, as the rubble created in its place could make a good hide or firing position for the insurgents to use against us but we knew this and the increase in range and field of view and fire was worth the trade-off. We could now study some of this new and unseen ground area in great detail with many hours of daily observation and hopefully notice any changes or differences because we had a small advantage with a slight gain in elevation from our position in the Sanger.

The attack helicopters soon moved off as a pair back in the direction whence they had come and out of sight to us. We waited and watched the area and eventually stood down, then went into our usual drill. The Sanger had to be cleared of ejected empty cases and link and then replenished with ammunition, and the used GPMG was taken away to be cleaned and replaced by another, oiled up and ready to go. Finally a verbal back-briefing or SITREP was given to the boss on what had just taken place. The sentries were changed over and went back to providing security of the patrol base as it settled back down and prepared to move into nighttime routine with last light rapidly approaching.

As for Matt and myself, at some point during the night we both like everyone else in the base had a stag: a duty for me as watch keeper in the operations room, and Matt on sentry out in the Sangers. It had been an eventful day. Nevertheless, it was one more day down, one more day closer to going home for all of us.

Chapter Ten

Patrol Base Silab

Part II, Overwatch

Our boss Lt D had delivered his orders to us earlier the previous evening outside the small operations room. All those who were to be involved were gathered around a 6ft improvised wooden table in the centre of a small open courtyard completely surrounded by a 6ft wall and this was the central command position of our patrol base, the heart of the location. The boss covered our upcoming mission in great detail and how he perceived we were going to achieve it, using the relevant mapping spread out over the table as an aid to his briefing as we all listened intently to his plan. Our primary function was to be in a position out on the ground where we could cover a large area with observation and fire if needed while a dismounted rifle company and its assets were conducting operations. This would commence with a company advance to contact; then once contact was made the ground and compounds around the area of Shin Kalay were to be systematically cleared of insurgents. The operation would be conducted by 2 PWRR.

That evening after the boss's orders were completed, we spent a short time finalizing the preparation of our kit and equipment to be sure that we had exactly what was needed for the task. The only rehearsals that we had to conduct were the packing, debussing and setting-up of the specialist equipment from the backs of the vehicles in total darkness and silence so that by H-hour we would be in our chosen position out on the ground, ready and waiting for first light the following day. The only issue was trying to squeeze everyone and everything required into the backs of the wagons. Making space for the additional manpower needed for the task and all the extra spare ammunition boxes that we could accommodate in and around the HMG tripods and guns and Javelin equipment was proving to be a bit of a drama but eventually we managed to fit it all in.

There was a lot of other essential equipment for our tasking, such as spare batteries for the radios and ECM equipment, litres and litres of drinking water, spare rations, medical equipment, and a few piles of empty sandbags plus some large picks and shovels for digging in if necessary. Or the sandbags could be filled with earth from the rear of our location out on the ground to build up our position, improving our protection from any incoming small-arms fire and fragmentation from either RPG or mortar fire. We would also need some loaded sandbags, either full or half full, to weigh down the end of the tripod legs when the guns were set up in the SF role. Doing this would make them a little more stable and improve their accuracy when firing in the sustained fire role at a rapid rate of burst fire, especially when engaging targets out at further ranges.

We could potentially be out on task for some time, depending on what happened out on the ground as the day progressed and as the tactical situation and threat changed. We could also possibly be re-tasked to another role and had to be prepared for this. Our primary task once again was to provide overwatch in a static location: waiting, watching, listening, reporting and taking action when needed to destroy or neutralize any insurgent threat to our comrades and ourselves, once that threat was located and PID. Throughout our time on task the vehicles would remain with us, the crews providing flank and rear protection while we were in our location observing and covering forward into the area of interest. So if we needed to become mobile at any point we had the option immediately available to us of being able to pack up our position and mount straight back up into the wagons and redeploy where and when needed without having to wait for another call sign to pick us up.

The passage of timely and accurate information on an all-informed radio net is just one of the key elements in any phase of any operation and is essential to commanders on the ground. The situation is always changing or evolving and any information on the enemy is as vitally important as bullets and bombs raining down on them. Therefore before deploying a 100 per cent check of the radios and associated equipment was always conducted, ensuring that everything was functioning correctly and also that our back-up communications plan would work.

It was a very early start for the whole patrol base that morning as the majority of the platoon would be involved in this task and would be deploying out on the ground with the boss. The few men not deploying out that day would remain in the base to maintain security by acting as a guard force and manning the Sangers while we were out on task. Moggy would be manning the small operations room along with a runner and Stu would be standing by with his No 2, ready to man and fire the 60mm mortar. Over on the gun line of the 105mm light field guns their whole set-up was crewed by the Royal Artillery and a few gunners attached to them from the Australian army who were all getting themselves and the guns ready for action.

We arrived at our proposed position after being mobile for forty minutes or so from leaving the base. The vehicles moved slowly and cautiously through the early-morning darkness in single file with no lights on and the drivers using their night-viewing devices to negotiate the gloom. First light was starting to break on the horizon and we could just make out the slight colour change from the cold darkness of the night as the dawn of a new day began to break through on the skyline with the sun starting its slow climb into the early-morning sky. Light was slowly penetrating through the dark clouds that blanketed the heavens, breaking through and beaming down onto the earth below. As I observed this effect it reminded me of the images associated with a classic old big-budget biblical Hollywood epic from the early '50s or '60s, just missing the big music score to accompany the spectacle; the beginning of a new day on the ground in Afghanistan. Gradually the darkness was replaced by bright colour: a mixture of light purple and red that tainted the low-lying dark clouds and skyline on the far horizon.

The vehicles remained in single file for the whole route as the small convoy snaked its way through the darkness and on occasions the complete call sign had to come to a halt to conduct VP drills. Once the obstacle was cleared and the dismounts had mounted back up into the vehicles, we were then good to continue on our way. The lead vehicle commanded by the boss came to a slow rolling halt just short of the proposed drop-off point and the remainder of the convoy halted in turn. The boss and his dismounts went forward and checked the drop-off point and once he was happy that it was clear and was the right location, he signalled for the lead vehicle to come in to him and as it moved slowly forward, the other vehicles followed on behind in file to occupy the DOP.

The vehicles only moved a very short distance and then stopped again, leaving just enough room in-between each wagon for the dismounts to be able to get out of the rear and unload all the kit and equipment needed for the operation. The top cover sentries in the rear of the wagons remained where they were, providing protection and scanning the area through their night-viewing devices. It was eerily silent all around us and hardly anything stirred; the early-morning quiet was broken only occasionally by the howling and barking of some wild dogs running around somewhere over in the middle distance.

We dismounted and quietly sorted ourselves out with the kit and equipment, then went about our task of moving off from the vehicles and clearing our way down to the pre-selected position on the ground that we were going to occupy and set up our MSG position, checking for IEDs and any recent insurgent ground sign as we moved. We were in a very hostile and active area with many recent contacts against the insurgents, so every one of us was currently in a heightened state of alertness.

Every time we halted with all the extra weight that we were carrying down to the position, I could feel my old, aching, sweating body asking my brain: 'What the fuck are you doing? Why are you still doing this?' My forehead and the hair underneath my helmet were constantly damp with sweat, slowly dripping small beads of salty-tasting moisture down the sides of my face and onto my collar. My shirt around my armpits and forearms was damp, sticking the material to my skin in places. There was a constant sensation of sweat dripping down the centre of my back under my body armour, trickling down towards the waistline of my trousers and making my belt area damp. All the while my mind was in overdrive as we moved cautiously through the darkness, thinking of the increased threat from IEDs that the insurgents were starting to lay everywhere and hoping in the back of my mind that the instant bright flash and sound of an explosion from such a device would never happen.

The area of ground that we wanted to occupy as our main position had been bypassed before on foot and mobile patrols and we knew the general area very well, aware that this location could offer us good fields of view and fire over our main area of interest. It was a piece of ground in the centre of a large open area in-between some single-storey buildings that were occupied by the locals and their families. There were some smaller outbuildings used to house animals and these were closely sited around the main inhabited compound buildings. The outbuildings were used as family stores away from their main living area and housed either chickens or turkeys and a few goats, with most of the floor space being taken up by long dried grass piled up in bundles and neatly stacked against the back walls. All the buildings were solid with thick walls and sturdy well-made flat roofs with only two or three window apertures in each individual building and one main entrance.

The open ground between the two main compounds was firm underfoot and had a low-lying wall that ran the entire length in-between the two main occupied buildings and continued along going away from us on our left where it then met another similar wall coming up from the bottom of the road, forming a prominent T-junction where the two features joined.

Over to our right flank were several further similar-sized outbuildings that housed some more resting animals and stored crops. Nearby was a prominent sturdy-looking water pump and littered on the ground around the base of this was what appeared to be an assortment of metal tins and brightly-coloured plastic containers. Behind and beyond these buildings for as far as the eye could see, it looked in the poor light as if this area was literally covered with boulders and small rock formations of all sizes lying around everywhere, and we could just make out the remnants of old compound buildings and their outer walls that had been left to ruin or damaged a long time ago and remained unrepaired and uninhabited. The ground kept its elevation, looking back down towards the road as it contoured round, following the main road and canal below back in the general direction of our patrol base, and about 450 metres further along from the base was a prominent crossing-point across the canal: a VP (vulnerable point).

What we could observe out to our front was that the ground very gradually sloped down away from us towards the canal and the road, and the distance between the low-lying wall that we would be using and these features was just over 200 metres. The road was approximately 15 metres in width and consisted of hard, dry, uneven, compacted earth covered in potholes with an irrigation ditch on the far side running along the entire length as far as we could see. This road was a main transit route used by the locals either on foot, in animal-drawn carts or motor vehicles to travel between the main towns and smaller villages and it was busy at times. The coalition forces also frequently transited along this route for re-supply to the patrol bases and for patrolling.

We were all in and good to go and were settling down into our new position and familiarizing ourselves with our new surroundings: the ground and the general area forward of the low-lying wall. The boss had ensured that he had good communications to the relevant call signs out on the ground and with our operations room. The two lads from the Royal Artillery who were our FST also ensured that they had good comms to the gun line back in Patrol Base Silab and with Camp Bastion for calling in other assets, always remembering the old adage of 'no comms, no bombs'.

Spenny and Joseph set up the GPMG and tripod first in the SF role, then loaded and made the gun ready to fire so all they had to do once we were engaged by the insurgents or had seen any insurgent activity was to PID the target or threat to us or other friendly troops out on the ground. Once the gun was aligned onto the target with the correct range set on the sight, they would remove the safety catch and fire, starting with one or two short-ranging bursts, then engaging the target with longer deliberate burst fire or rapid rate of fire, depending on the requirement to neutralize the threat.

They also prepared the remainder of the 30mm belted ammunition for use from the newly-opened H83 container and piled up a few of these brown containers by the base of the wall directly behind the gun for the two of them to sit on so they could observe out over the wall and operate the gun comfortably. The Javelin post was sorted out by Big Sam and set up ready for use, ensuring that he could observe and fire from this position. He had three spare missiles with him which were set down at the base of the wall to the right of his firing position; again close at hand should he need to fire again rather rapidly.

While all this activity was going on Matt and I went firm, putting ourselves, our day sacks and two H83 containers of ammunition – one of .338 ball ammo and one of 7.62mm ball – down between the SF gun and the Javelin and set ourselves up there behind the wall. All we had to do was put our rifles down after a quick initial scan out over the ground through our optics and then do the same using our binoculars. We also made use of our spotting scope to observe the ground in more detail and the areas surrounding and leading up to a few of the larger compounds located out to our front at varying distances. Any prominent physical features were identified, either man-made or natural; basically anything that we could use as a quick reference point in a given arc of observation to rapidly get ourselves onto a given target no matter where it was or how far away from us. We made good use of the laser range-finder to ensure the accuracy of this potential target information and plotted it onto our range card for future reference should we need it.

Meanwhile the wagons that had dropped us off had moved away slightly to occupy individual positions around us providing flank and rear protection while we covered out to our front, meaning we had all-round defence for our complete call sign on the ground. The wagon just slightly over to our right was a WMIK and that vehicle was commanded by Bez, a big South African, and mounted on the rear was the 40mm grenade-launcher. So in all we had some firepower and even greater firepower on call if we needed it. Everything was now set: the complete call sign was firm on the ground and observing out over their respective arcs. The fact that everything had gone according to plan was an added bonus as more often than not when we left our location on a mobile patrol, either by day or by night, it didn't take long for us to be taken on by the insurgents at some point.

We were all settled down in our location and silence reigned over our small defensive position. We were good to go, this was it: now all we had to do was observe and wait. No matter what the forthcoming day had in store for us, we were all ready and waiting. The dawn of yet another hot and dusty day soon replaced first light and was now upon us in all its glory: the darkness overhead had been replaced by a clearer, much brighter sky compared to the early-morning low cloud cover of several hours earlier.

There were just a few large whitish clouds seeming to hang motionless high above us and dispersed widely over the sky for as far as we could see. Visibility all round and out to the middle and far distances was very good with the natural eye and our sight picture and detail when observing through our optics was crystal clear. While conditions remained cool there was no mirage or heat haze affecting observation out to the middle and far distances, although this could possibly cause distortion at further ranges. At the time the ground was still cool from the night and had not yet been warmed by the sun. It was a good start to the day: almost perfect light and wind conditions for our task.

Everyone who had optics had adopted the kneeling position or a variation of it to observe and was peering over the low wall, using the wall itself as a support for their elbows, scanning the ground out to their immediate fronts and taking notes; also marking the respective range cards within their groupings and lasing prominent points or reference points out on the ground and marking down possible targets. Matt and I were doing exactly the same, constructing a detailed range card and constantly assessing the wind direction and strength. As usual we broke the ground area down, giving ourselves a left and right of arc and an axis dividing the ground in two then breaking it down further into the near, middle and far distances. A number of prominent reference points such as track junctions, large compounds or similar features within these smaller areas of observation were clearly identifiable and could be used for quick target identification.

We were smack bang in the middle using this low-lying wall and piece of open ground between the main buildings and smaller outhouses that were more or less either side of us and slightly offset from our position. The buildings were roughly 70 metres away from us and the stationary vehicles on the flanks had these locations covered by view and by fire if needed. The dwellings were ground-level buildings with thick hard-baked mud walls; they were also inhabited and the locals from these buildings started to come out. Men, women and children of various ages emerged from the main entrances to their homes and stood there in front of them in small groups, just silent and staring at us for some time. It was only the children and the dogs that made some noise and they were soon silenced and restrained from coming over to us.

I think the adult males were probably just in total disbelief or shock. As they stared at us, the hatred could be seen building up in their dark piercing eyes from below the long dark hair under the front of their Talib headdresses. Each and every man was heavily bearded with a weathered darkened face. The atmosphere and tension on both sides was tangible: mutual suspicion and distrust manifested in total silence.

Their whole manner seemed to show complete contempt: a loathing and sheer disgust towards us, judging by their eye contact, facial expressions and overall body language. In private they were surely thinking: 'Why are these western infidels here? How dare they come onto our land, suddenly appearing from the darkness and occupying themselves down here?' However, in my opinion it was rather more sinister and I was not the only person to feel this way. All the males were of fighting age and all were dressed in very similar clothing and in the same location at the same time. Maybe this had been their hide or lie-up position from the previous night? One of the men certainly stood out from the remainder of the group as their possible leader; he was definitely older. Recalling recent events in this area with engagements against us from the insurgents and the type of incidents, just looking at this group suggested they had the able-bodied manpower to do the job; they appeared to be just over a section of men in numbers.

Minutes passed by, still in total silence, with both sides still watching each other. Sure enough, after a short period of time the presumed leader made a gesture towards one of the much older women within the group who stood in front of the other females and the children who were now all more or less grouped together. Several moments later all the women and children started to move away, going in the general direction over to our right in a group and after a couple more minutes of waiting and watching us, so did all the males who began to move off in pairs and walk towards the T-junction of the wall to our left. Occasionally they looked back towards us as they moved behind the wall and out of sight. All that was now left in this small area was just us and a few large animals, and they were only there because they were tied up and therefore unable to move away.

Any soldier will tell you that this was not a good combat indicator at all, the classic indication that something bad was probably going to happen and happen very soon, and everyone else in the call sign had picked up on this as well. Even as the minutes were slowly passing by we noticed that the road seemed to quieten down until it was free of people and traffic; in fact the whole area out to our front seemed to be clear of all human movement.

We readied ourselves and waited for something to happen, observing out to our fronts with each man behind his individual weapon system and now keeping a much lower silhouette than usual, using the cover of the wall and just waiting and watching, poised ready to fire. Everyone was scanning cautiously over the ground using their rifle optics or binoculars and our

Mark I Eyeballs for any sign or any indication; just a glimmer of what was about to happen. Maybe we would catch sight of the insurgents manoeuvring around over the ground out to our front or flanks, trying to move into their firing positions. Our radios went silent for a short while. All that could be heard was some static interference every now and then or a short, sharp, hastily-spoken message for another call sign somewhere out on the ground, an unfamiliar voice suddenly coming from the small black handset, then silence again.

Everyone was focused, alert and poised; knowing that at any moment the shit was going to hit the fan and we would all be engaged in fighting the insurgents once again. It is the waiting and uncertainty that puts you on edge, the anticipation of something that you know is coming: the mind races and the heart pounds, speculating as to when, where, how many and with what are they going to take you on? We were constantly and instinctively trying to assess the threat in our heads, knowing the reality of the danger and going into fight and survival mode. After a second or two we would react and try to counter-measure the threat and then fight back, using our gut instinct and training.

All we knew for certain at the back of our minds was that when it all kicked off, we *must* react. The rifle or LMG rounds would start ripping up the ground around us, slicing through the air above our heads and around us, an RPG or two would suddenly come from out of nowhere, winging its way towards our position and exploding in the air or on the ground, mortars might start landing around us or we may even receive a casualty from within the call sign. Within an instant of all this happening we would be in real deep shit; immediately engulfed by intense piercing noise, sheer violence and chaos surrounding us and our senses all working overtime. The insurgent's sole intention is for us to die. But in my mind: 'Not fucking today, sunshine.'

Then it all happened: the silence was broken and the waiting was finally over. The insurgents had initiated the contact, opening up their attack with small-arms fire and the first rounds came zipping erratically towards our location, coming from somewhere out in the low ground to our front. 'Contact front!' was shouted out. Then our whole world erupted with instant noise and hot lead all in a few short moments. To begin with the insurgent rounds were going high above our heads but it did not take long for them to realize this and adjust their fire accordingly; soon the enemy fire was starting to find its mark on and around our location. The incoming fire started to fall just short and strike the ground out to our front, leaving small dust signatures as the rounds impacted the hard earth. A number of rounds managed to find their mark in the wall in front of us, striking its earthen surface and spitting up small pieces of dried mud into the air, making a snapping sound with every hit.

The incoming fire was initially slow but then the rate from small arms seemed to pick up momentum and intensity, causing everyone to keep as low as possible. We all tried frantically to scan the ground from whatever firing position we had adopted, using all cover available in trying to reduce the chances of getting hit but continually trying to positively identify the source of the insurgent fire and put down some accurate return fire towards their positions.

Then Spenny's GPMG joined in the chorus, spitting out ball and tracer at a rapid rate of burst fire and sending 7.62mm back across the open ground towards the road, slicing into the forward edge of the field on the far side, cutting into and through the long grass and smacking into the front edge of the ditch, kicking up mud and plant debris. This field had a long, thick, very green crop of some kind growing densely compacted and the forward edge of the field came right up to and ran parallel with the road with a frontage of just over 150 metres in length, our imaginary axis cutting straight through it.

Looking at the source of the insurgent fire we had a slight advantage due to the elevation of the ground so we were actually viewing slightly down onto this area and the insurgents would have to fight their way up towards us. The field that they were using for cover from view, and possibly the ditches for cover from fire, went back away from us for roughly an additional 250 metres from the forward edge of the road. We had to try to observe into this large green mass as at this point all the insurgent fire seemed to be coming from this area directly out to our fronts. They must have used the cover of the low-lying wall and the rear of the field to mask their movement and screen themselves, allowing them to move into position without being observed from our location.

Suddenly the insurgents were on the move through the thick vegetation and coming forward towards the main road to our front at a rapid pace; the foliage was being disturbed, indicating where they might eventually emerge into view. Spenny was still firing the GPMG at a rapid rate of burst fire, while Joseph acted as the gun's No 2 loading a fresh belt of ammunition when needed and helping to feed the belted ammunition out from the H83 container and into the gun. He also dealt with the spare barrel changes, helping to reduce any unnecessary stoppages associated with prolonged firing of the gun. Individuals along our line to my left and right started to fire their rifles; empty cases were being ejected and starting to collect on the floor around us. Matt and I were very close together, almost side by side, in a variation of the kneeling position with both knees firmly and evenly placed on the ground as we leant slightly forward using the wall as a support. The top of the wall made a good rifle rest where my bipod supported the rifle while the butt was firmly in my shoulder.

I was scanning the forward edge and sides of the field through my rifle scope on x12 magnification, while Matt was observing through the binoculars and doing exactly the same. I was able to make minor adjustments to my position and observe out across the near and middle distances overlooking the ground to the left and right of our axis. The rifle bipod was set at the correct height for me while in this firing position and by raising myself slightly off my boot heels or lowering myself by bringing my arse closer to the heels and flattening my feet I was able to adjust my sight picture by moving it up or down slightly as necessary. The noise around us now was horrendous, even while wearing ear defence. I looked away from my scope for a split second because of the sheer volume of noise coming from the GMPG that was firing to the left of me and glanced towards the end of the GPMG barrel: there was an almost continuous small, bright, dirty yellowish flame coming from the barrel through the flash eliminator and the sides.

I then turned my head back so my face was in line and to the rear of my rifle scope, rested my cheek back down on the adjustable cheek piece and went back to observing, adjusting the scope magnification from x12 up to x16 and then adjusting the fine focus of my sight picture for absolute clarity. I was concentrating on an area of ground that was the forward edge of the centre of field directly to my front where it met the road. I will never forget this moment and still can't figure it out today: either it was ultimate bad luck for the insurgent, or fate, destiny, synchronicity or whatever you want to call it. But it happened right in front of me: I was in a firing position looking into an area and ready to fire when needed, simply by looking away from my rifle scope in this fixed position and then moving my head back again, all without having to actively seek out a hostile target to engage. An insurgent standing fully upright and moving forward just instantly appeared and filled my rifle scope.

I was observing through the scope, my safety catch was off and I was ready to fire; the tip of my right forefinger just touching the trigger. Then at the exact pinpoint location where I was observing, this upright figure came crashing out of the dense undergrowth clutching an AK variant in both hands at chest height. As he forced his way through the screen of green foliage, broke free of the field and came out into the open, he was literally directly to my front and facing me, bringing his weapon to bear at waist-level with a look of sheer aggression on his face. It seemed he was about to come suddenly charging forward towards us firing with bayonet fixed. His whole upper torso now filled my sight picture and from the absolute clarity of the image it seemed like he could just reach out and touch me.

My brain registered this information in a flash: both the image and the threat. Matt had also just identified him observing through his binoculars and went to give me a target indication but I was already on it. My mind and my thoughts went clear as I focused on the insurgent and what I was about to do. After only a very minor adjustment to my point of aim and firing position, I superimposed my crosshairs onto the centre of mass of the target which was the upper chest area. Holding that sight picture, I let out a slow, controlled, deliberate breath as I operated the trigger, slowly and deliberately increasing a little pressure through my forefinger. Then at a given point of the trigger's smooth rearward movement an 8.59mm ball round was released and sent towards the intended target.

The next moment the round impacted into the target's body, momentarily creating a small blood and cloth spatter that came off at my point of aim on the upper chest, instantly sending him violently and forcibly backwards into the thick green foliage of the field, more or less where he had come from. It all happened so quickly. One second he was there and the next he was gone: target down.

As usual the next few minutes were vitally spent in observation. I continued scanning the area through my rifle scope while simultaneously operating the bolt handle using my right thumb and forefinger, bringing it fully to the rear to eject the empty case from the chamber. I then pushed it back smoothly, going fully forward and picking up and feeding into the chamber another 8.59mm ball round from the magazine so I was now ready to fire again. No movement was to be seen at all: neither by man nor the natural foliage in my area of observation. Sometimes movement of foliage may be attributed to man, or to the local wildlife that inhabits the area. Either way, once you have eliminated one or two other factors it can alert you that something is there: another form of combat indicator known simply as telegraphing.

Several moments passed by in my own little world, much as I tried to block out everything else going on around me and concentrate solely on what I had to do. The noise around me seemed to intensify as I broke from my position to have a quick scan around to my left and right and scanned the area back out to our front but it seemed clear. Then as quickly as it had all started, the firing came to a halt but everyone remained highly cautious and alert at their positions, still keeping a low profile while scanning the ground through their rifle optics. Some now needed to do a quick magazine change so they were ready with fresh ammo should it all kick off again. The boss had his map out in front of him and his radio handset in the other hand and was trying to send his SITREP through.

Soon after this on the left-hand side of the field as we looked down a small group of insurgents suddenly broke from the cover of the forward edge of the field and started to fire, then fire and manoeuvre coming towards us for a few seconds, finally disappearing out of sight behind the cover of a low-lying wall. Over on the far right-hand side of the field, again on the forward edge, another group opened up on us. Our immediate response was to put down a rapid rate of fire back towards that area. The 40mm GMG covering the area soon started firing at a rapid rate and as the gun fired we could see the vehicle rocking. A good few well-placed HE rounds from the 40mm well and truly decimated that small area of ground. Clumps of brown earth and green foliage were thrown up several metres into the air with each impact; the detonation of the HE rounds finding their mark leaving just a greyish cloud lingering in the air once the debris had fallen back to earth. Soon that would be dispersed by the wind and the threat had been silenced.

Back over on our left side the insurgents were still firing and had started to move up on our left flank using the cover of the wall that came up towards our location from the bottom of the road. This gave them, or so they must have thought, some cover from view and fire and it did to a degree. As they started to move up in single file every now and then a head and top half of a torso was momentarily visible; then they were gone and reappearing somewhere further along the section of wall coming towards us. Spenny and Joseph rapidly switched the GPMG as far right as they could get it and started firing short bursts in the direction of the wall. We could see from our location that the vehicle crew providing flank protection over on that side was engaging the insurgents who must have been by now more or less directly out to their front. This was with a mixture of rapid rate of rifle, small-arms and LMG fire from the Minimi gunner being fired from the top cover position in the stationary vehicle: it looked like a continuous stream of red tracer and ball that finally brought their attempt to outflank us to a decisive end.

Our priority, now that we had dealt with defending our small position and the threat had been successfully stopped, was going back to concentrating on our main task of providing overwatch for the dismounted rifle company out on the ground. One or two call signs from A Company had started to note suspicious activity within their areas of operation and the radio net was busy with call signs passing along this information to each other. The boss had the handset of his radio more or less permanently pressed to his ear, relaying information as to what the company was doing and where they were located out on the ground whenever the information seemed vital for us to know.

As the sun continued to climb high into the mid-morning sky, the air and ground temperatures began to rise and for everyone, especially the dismounted rifle company, it was going to be one hell of a hot day with many hours of ground-fighting to come. The cloud cover was pretty non-existent: just a clear blue sky for as far as the eye could see. Time was passing by and more or less everyone was still in their same positions from first light, either kneeling or sitting and continuing to observe out over the ground.

The majority of the brass and link from our previous engagement had been cleaned up off the ground in and around our position, sandbagged and then stashed away on the vehicle. It could not have been any more than an hour or so that we had spent in relative silence, just observing, and throughout this time there was hardly any movement of the locals or of motor vehicles transiting through or around the area. This was highly unusual for the time of day, so again it was a case of watching and waiting to see when and where the insurgents would next strike.

High above us in the sky a pair of Apache attack helicopters was now on station operating in our area and prowling overhead. Something was happening over on the far side of the town as we viewed it from some distance away: whatever was happening was out of sight to us at the moment but had caught the attention of the Apache pilots. Everything seemed to be starting over in this small area of ground. The faint sound of small-arms fire and dull muffled explosions way off in the far distance could just be made out. On hearing this and as the noise started to grow louder and was clearly identifiable we tried to get eyes onto the area from which the sound was coming.

The boss's radio sprang into life once more, as from where Matt and I were sitting even I could clearly make out the short, sharp voice transmissions coming from the handset. The rifle company troops on the ground were now in contact in the area over to the far side of the town from where we were currently observing. We could just make out through our optics some friendly forces call signs moving around, involved in what appeared to be close fighting around a few of the compounds on the far outskirts. All we could do at this point was report what, if any, insurgent movement we were able to observe.

We started to notice the activity of several motorbikes cutting about the area at speed some way off in the far distance and out on the flanks of the large area that we were observing. These could be seen moving from compound to compound, suddenly disappearing out of our sight, then reappearing after a few minutes and travelling to the next compound or making their way towards the outskirts of town using the dirt tracks or going cross-country. Once they had made it into the built-up area of the town they disappeared from view again but after some time would reappear elsewhere. Matt and

I would mentally log the types of motorbike and the riders' clothing, any distinguishing features, so we could trigger these individuals around the area.

From past experience we knew the insurgents used motorbikes and other vehicles to move their commanders around the fighting on the ground so when things were going wrong for them they could easily extract high-value commanders by quickly getting them out of the fighting and away to a safe house to plan and conduct future operations against ISAF forces. Foot soldiers are replaceable; experienced battle-proven commanders are not. Motorbikes were also a means of passing on vital information to insurgent fighting formations on the ground when quite possibly their radio and mobile phone communications were poor at the time.

The noise of the fighting intensified and was now clearly audible: the sound of single-shot small-arms fire mixed in with long bursts on either LMG or HMG and the detonations from either the 40mm or UGL could be heard. Plumes of black smoke could be seen coming up from various compounds or ground areas where the fighting was taking place. There had been a few large explosions and we presumed these might have been the entry teams gaining access into the compound as no CASREPs or reports of IEDs had been sent over the radio following the detonations. Our FST was soon on it and called in a fire mission onto an area where we had spotted some insurgent activity. I remember sitting there observing through my rifle scope waiting for the artillery to rain down and when it did, all I can say is that it was truly impressive to watch and that you cannot argue with a 105mm HE round when caught out in the open or in a confined space.

Everyone in our MSG and protection group was involved in some way, doing their small bit in support of the rifle company who had by far the hardest role in having to fight through and clear the compounds and open areas leading up to them for this small operation to be a success at the end of the day. We remained in our location at our positions until last light and when the fighting finally came to an end we stayed in location until the company had completely extracted off the ground to cover them. Then we finally cleared up the location, as always trying to remove as much ground sign as possible, then simply mounted back up into our vehicles and made our way back to the patrol base without incident.

Big Sam who operated the Javelin fired two missiles against insurgent targets that day and had two cracking confirmed strikes against them. As for Spenny and Joseph, they would be spending the next few days cleaning their beloved GPMG and the spare barrels and explaining to the CQMS why they needed to exchange a barrel or two on the next re-supply due to all the excessive and prolonged firing that they were involved in on that day, also that the gas parts complete on one of the barrels nearly went down range along with the rounds on one occasion. That day on task providing overwatch I would compare to some of the fighting that we were involved in on Operation SOND CHARA and the long days and nights involved in that operation for the ferocity and intensity of the fighting on the ground at times. It was the most insurgent activity that we had seen and been actively engaged by for some time with regard to the insurgent strength in numbers and to the kit and equipment available to them and that was used against us. However, once again we were very fortunate and did not receive any casualties during that day.

As for myself and Matt, during the early morning and through into the middle of the afternoon we were involved in two more engagements as a sniper pair and from a sniping point of view the first of the two was a straightforward task. The overall range to the target from our location is estimated by measuring the target using its known height, times that by 1,000 and then dividing that number by the height in millimetres and is a good method to practise if you have the time to use this method. But more often than not you are about to make contact or are already in contact being engaged by a hostile threat and the need for speed and accuracy in neutralizing that threat as soon as possible is your priority. Therefore the laser range-finder is used first and foremost and is in almost constant use by a sniper pair.

After using the laser range-finder for final confirmation and accuracy to ensure that my range data was as precise as possible, the range to the target was lased at 810 metres. The target was now static and seen to be making use of some form of hand-held radio or walkie-talkie with a narrow, dark, whip-type antenna. This he stuck outside his vehicle's side-door window while remaining seated in the vehicle talking into the radio, and by his whole demeanour he seemed to stand out from the remainder of the group. In our opinion he was the leader, in command of this small band of insurgents.

There were now three insurgents standing around the front and two seated in the open back of this pick-up vehicle. Both side-door windows to the front were fully open, meaning my round would not have to penetrate through any glass as this could affect shot placement onto my point of aim. My target was seated behind the steering wheel of this banged-up white 4x4 vehicle that we had previously spotted and had been following it, triggering the vehicle and its occupants around the area because it carried several armed insurgents. But the target was difficult to take out or engage with accurate and decisive fire at the time due to other factors and we wanted to make sure we got all of them in one definitive hit. The vehicle had been appearing and disappearing, in and out of cover from view and fire, by making good use of the compounds and the undulating terrain of the ground to screen them as and when they went mobile.

Just as we were going to call the Apache helicopters in to do the job, by sheer luck once again the vehicle appeared in our field of view and stopped beside a compound out in the middle distance, well within our arcs of observation and fire. There was no time to waste: this was a once-in-a-lifetime opportunity to take out such a priority target and immediate action was needed. Without further delay I aligned myself and my sight picture onto the target and adjusted my focus, choosing my point of aim as quickly as I could.

Then I just simply squeezed the trigger gently and fired my rifle, and in a moment the round impacted into the side of the target's head as it appeared to me through my rifle scope. The top of the upper skull appeared to rise up, momentarily coming away from the top of the ear area, then came back down. This all happened in a fraction of a second. The head then slumped down towards the chest, followed by the head and upper torso together slowly sagging forward, leaning on the steering wheel and dashboard. The vehicle started to move slowly, rolling forward slightly, coming off the track and rolling into an irrigation ditch. Almost immediately after I fired, the GPMG and 40mm GMG opened up and took care of the remainder. The wind error was minimal and so was the mirage affecting my observation of the target area. After just a few short minutes of both sides firing frantically at each other it was all over and that particular threat was now gone.

For the second target the range from our location was finally measured at 1,544 metres using the laser range-finder mounted onto a small tripod to aid accuracy at this distance. This was not as straightforward to engage as the previous target as the overall distance had vastly increased and so would the error if we got our firing calculations wrong in estimating distance and wind corrections plus one or two other environmental factors that can affect the placement of a round onto a target. The mirage as we estimated at the target end was drifting at an oblique angle and the wind, even out at this range, was moving across the open ground yet going towards the target at the same angle. Based on our very quick observations of this effect of wind and mirage, it was surprisingly very minimal at this distance as I determined through my rifle scope and Matt with the spotting scope.

It took us two rounds to get onto the target at this range and then the third round hit and the target went down. The first round I fired we did not see the fall of shot or even the swirl that can sometimes be observed when firing as the round travels, displacing the surrounding air as the projectile moves at great speed creating a swirling, tunnelling effect. On occasions this can be observed through an optical scope in the right environmental conditions.

The round seemed to have dropped very short, so we must have wrongly estimated the range to the target. I immediately increased the range setting with an alteration to my elevation/range drum and we both momentarily double-checked the wind corrections I had put on my deflection drum. The target was still firm but time was running out: he would soon be on the move and would stop firing. I picked my point of aim at the centre of observed mass and fired again. This time the fall of shot was observed by both of us and the round dropped just a little short in the foreground before the target. Luckily for us, with everything else that was going on around the target and the fire-fight he was involved in, he did not notice this as we were trying to range in.

There was not a minute to lose and once again I rapidly chambered another round ready to fire. This was my third attempt; I adjusted my range drum accordingly and was happy with the wind setting on my deflection drum. I concentrated and focused hard as I found my point of aim again, held that position, steadied my breathing and fired my third round. What felt like whole minutes to me passed by while I was still observing the target through my rifle scope and thinking: 'What if he fucking moves in the next second or two, or I miss again?' All the time I was bearing in mind the actual distance to the target and the time it would take for the round to cover that distance. Then finally after its long journey the round made contact and the target fell to the ground. Through my rifle scope it looked like someone just stumbling or tripping over and falling, but this time there was no getting up.

The target had been in the standing position in front of a large compound building that was his backdrop and within a few metres of his motorbike that was supported by its stand near a low wall with some tall, thin deciduous trees on the other side. He was in a firing position and facing towards the direction of a call sign that had just been caught out in the open ground as they moved towards a compound on the outskirts of the town. I remember just watching that scene for several minutes through my scope: the very bright red colour of the fuel tank of the motorbike and the black seat that extended over the rear of its silver mudguards. This image still sticks in my mind, even today.

After some minutes had gone by I broke from my firing position to use the spotting scope to observe into this area. There was a lot of activity both from the friendly forces and the insurgents as by now it appeared that the friendly call sign was having to fight forward to get out of the open area of the large field and into the cover of some small outbuildings on the forward edge of the village.

The location where the insurgent had been standing and firing was almost right beside the compound building that must have contained more insurgents who were using this location to open up and engage the friendly troops from within while they were crossing the open ground of the ploughed field. Within a few minutes this area was hit by a fire mission coming from the 105mm light field guns manned by the Royal Artillery. A number of rounds came winging over from the direction of our patrol base and impacted into the ground, detonating just short of the intended target and sending dirt and smoke up into the air. This was easily observed, and in no time once the corrections were sent back to the gun line via the radio the next salvo came over and the rounds were on target. The debris from the building and the surrounding area was mixed with a greyish smoke that rose up off the ground where the round had just detonated and formed into plumes as this mixture was forced up into the air with each new detonation. For us watching this event it was like having front-row seats in a cinema and watching the latest blockbuster war film in super HD with surround sound; however, this was real-life destruction.

The hours of daylight passed by very quickly for all of us with constant activity of some description and there is always a priority that pops up and must be dealt with, and so it should be; that is your task. Sometimes as the hours slip by with everything that is going on around you it's easy to forget that you have not got any fluids down your neck for a while, that you are getting very dehydrated and that you now hardly ever go for a piss and when you do, it is dark yellow and smells bad. So whenever you can, you down some fluids and a bar of chocolate. It is these small things that sort you out and keep you ticking over until the task is complete.

It is only when you are back in the sanctuary and security of your patrol base with your comrades and dealing with personal administration like weapon-cleaning, the preparation of your kit and equipment and then yourself, that you are ready to go again. Then when you finally get chance to get your head down, your mind and body let you know what the fuck you have just put them through as you lie there in would-be silence with the battle noise still ringing in your ears. Surrounded by complete darkness, you finally try to close your eyes and rest but there is one more final battle: a battle against the still very active and conscious mind and aching body that only you as an individual can fight. All you need and want to do is simply shut your eyes and rest and make it all go away.

Chapter Eleven

Patrol Base Silab

Part III, Counter-Sniping

Put simply, counter-sniping is a very dangerous tasking and failure in your task can mean death either for you or your comrades. The essential requirements are:

- A very strong desire to live
- Self-discipline
- Luck
- Knowledge and an understanding of your enemy.

Single-shot engagements against the platoon while out on foot patrol and against the sentries in the Sanger locations around our patrol base had started to increase. A comrade from another platoon while on duty in a Sanger at another base not too far from our location was seriously wounded during an engagement against the insurgents, having received a near-fatal single GSW to the head.

It had been a good clear day weather-wise leading into the afternoon, the wind and light conditions were favourable and the time was approaching mid-afternoon moving towards 1430hrs. Matt was preparing a brew and I was doing our washing. Our trousers and socks from the previous day had got a little wet and caked in mud while out on a foot patrol around our location as we were investigating the irrigation ditches and other possible insurgent firing positions, paying particular attention to new areas of foliage that were starting to spring up in the surrounding areas of the base. The days were now getting much longer and the sun was providing the vegetation with many hours of natural light and this, combined with the water that was still in the irrigation ditches and the ground, meant that everything seemed to be growing at a rapid rate and starting to conceal some areas that we could previously observe into, narrowing our field of view.

Suddenly a runner from the ops room came hurtling round the corner, through the single entrance doorway into the open area of the courtyard and into our field of view, coming to a halt right in front of us and momentarily pausing to get his breath back. He then explained that Sanger One had just been engaged by a single round, that the round impacted close to where the sentry was standing and found its mark in one of the wooden beams on the corner of the Sanger that help to support the roof, providing overhead protection to the sentry.

The runner, having passed on the message, returned to the ops room. We made a quick plan and having decided which optics to take, elected to split up. I would go to Sanger One and Matt would go the MSG Sanger and we could both then observe the ground out to our respective fronts. Our arcs of observation and fire would overlap as these two Sangers were roughly in line and only separated by some 200 metres. In theory we could then both cover the ground out to the front and flanks of the Sangers and observe the general area from two different angles or fields of view with some added elevation overlooking the ground from our respective positions. My location would enable observation into some of the many compounds nearest Sanger One as it was this area that the insurgents had favoured lately when engaging the base and the foot patrols. Mobile patrols in the wagons and the CLPs en route to our location also tended to get bumped as the long convoys moved slowly through this area.

I reached Sanger One and approached up the small incline of rubble and HESCO stairs leading into the back with caution. I tried to shield my rifle from view by holding it in the centre of the stock and straight in line with the right side of my upright body so if the location was being observed, hopefully the sniper rifle would not be noticed from the outside as I climbed up and entered the Sanger.

I kept myself low, as was the sentry, and prepared my optics, observing out with caution as he briefed me verbally on recent events and recent local activity of the civilians within this area. For me there was a pattern forming all the time in which we all constantly observed for any changes in and around our location and while out on patrol. We were both observing out over the ground with caution as there were a few small walled and open compound buildings directly to our front no more than 100 metres away with some small alleyways and tracks criss-crossing in and around the area.

Some goats and cattle were freely roaming around here and there was a poor old donkey that looked in a fair state of health. A crude harness around its neck had a length of thick old rope attached that was secured to the ground by a large peg, and the sad animal just seemed to stand there with very little movement. Off to the right of our field of view on the outskirts of this small urban area and just under 250 metres away was an area of natural foliage: a small copse of mature deciduous trees of over 30 feet in height. A prominent irrigation ditch ran along its entire length, cutting through some farmers' fields and going behind and into this area of interest. The whole extent of the ditch was still very green and thick with grass and vegetation but the open fields that it cut through were freshly ploughed and just contained overturned wet mud.

Much closer to our location and directly to our front within 20 metres or so was a T-junction just slightly right of our location and this junction was very prominent on the corner of our patrol base. This track or road was in almost constant use and sometimes very busy being the main transit track, a road route for the locals and the people out in the surrounding areas of the base. It was also used by the coalition forces for the re-supply of our PB and for transiting through to other PB locations around this area and to local villages and small settlements for operations. This feature also ran parallel alongside a man-made purpose-built canal that came and went in the same direction as the track, running in both directions for as far as the eye could see.

The canal was about 25 metres in width and had concrete sides. It was spanned by a very narrow military-style constructed footbridge and we often used this to cross when conducting foot patrols as doing so meant that on occasions we could bypass a prominent VP further down the track. This crossing was constantly under the watchful eye of the sentry in Sanger One and from this location the GPMG could be utilized if needed in support of a call sign in trouble in the area and cover them as they extracted back towards and into our location. Also by using this footbridge we could get across the canal on foot as a complete call sign pretty sharpish in an emergency and move straight up towards the high ground and then occupy it if necessary in order to cover other friendly call signs in trouble instead of patrolling a further 350 metres down to a culvert crossing-point. The latter took time to clear and then cross and was often a contact point covered by small-arms fire from insurgents in the fields and small compounds overlooking it.

The water in the canal was fast-flowing, looked rather menacing and had caused dramas to a few of the larger and heavier vehicles used on CLP runs in the past. On trying to enter or leave our location at night and especially under contact it was a very tight turn either to the left or right to get out of the patrol base and you could end up going into the canal if you oversteered and misjudged the turn as one or two vehicles had done in the past.

Both of these man-made features passed along the entire front of our base where the main entrance was situated. This was the only vehicle entrance and exit point, at which the main gate literally opened up onto the main track outside. On the track directly outside our front gate was a chicaned vehicle check point manned by the ANA where they conducted checks and searches of people and vehicles at their discretion. They had a very good knowledge of the locals regarding who to stop, question and search if they were transiting through the area.

Across the canal on its far side the ground slowly and gradually increased in elevation and was the highest natural feature around us in this area. The ground levelled out evenly at the top and underfoot it was just a mixture of hard desert earth and rocks. There was a permanent and very old local graveyard sited up there which was over to the left as we looked up towards it from our location and this was surrounded by a low-lying rocky wall. There were a few pieces of brightly-coloured ragged material of varying sizes attached to long, thin wooden poles standing upright from the ground just like flagpoles dotted around the graveyard. These were clearly visible from the surrounding area and were probably used to mark the graves of important local people. But I would say these were probably useful to both sides in estimating the wind, especially when the insurgents were firing their mortars and rockets towards our location.

There were many possible positions in which to lie in wait and to observe and trigger or engage us from looking down onto the road or our crossings of the canal via the small footbridge or generally over any sections of our patrol base from this very rocky and light area of vegetation up on the high ground. I started by using my naked eye, and just as all soldiers do, I scanned for possible firing points going for the most obvious and working quickly through to the least obvious within my field of view. I then worked through possible locations on the principle of line of sight and clear arcs of fire: basically if they can see me, in theory I should be able to see them. Based on what the sentry told me, the shooter must be close by as the time from the sentry hearing the shot ring out to the moment of impact when the round penetrated the support beam was very short, making an almost instant snapping sound, or crack and thump as we call it.

I looked at where the round had penetrated the beam and back again, and everything seemed to come into line: an imaginary rough trajectory line appeared in my mind as I looked out over the ground to my front leading towards the small clump of trees and irrigation ditch. I scanned the area of the tree line, starting from the bottom and working my way up the trees to the tops, then studied each one individually, trying to make out its detail or discover any secrets that might be hiding in the foliage but this takes time.

* * *

Valuable lessons were learned in the not-too-distant past about the use of treetop snipers, mainly during the Second World War and also on occasions in more recent conflicts. The height advantage gained in the trees can offer a better field of view and fire into the general area of interest where an enemy might be advancing. Also trees can offer better concealment to the sniper than ground cover at a given time, especially in a densely forested area of considerable size. So trying to locate and PID such a threat can be a major problem and take some time.

However, this is not a very practical firing position and is not recommended or taught for a whole range of reasons, such as the route in and out from such a location and then trying to get up and down the thing without breaking your neck. Also in actually finding a stable firing position and having any alternative back-up positions your options are very limited. Cover from view may also be limited and with possibly none at all from fire, sustainability and survivability may be very low for the sniper once he is in such a location.

Treetop snipers would have to somehow strap themselves onto or into the tree for additional support, both when firing and observing. For instance, Japanese snipers in the Pacific theatre were known to have spikes attached to the front of their boots to aid climbing specifically for this task, making use of the tall palm trees growing on the Pacific islands. In the marshy and wooded expanses of land that covered huge areas of Eastern Europe, some valuable and harsh lessons were quickly learned by German forces in the invasion of Russia during the Second World War. Similar lessons were also learned on the other side of the world by the Allies fighting the Japanese in the jungles of the Pacific.

Trees had been used both by Soviet snipers in Eastern Europe and those from the Japanese Imperial Army in the Pacific, and were initially effective to the point of initial contact. However, once discovered there was usually only one outcome for the treetop snipers and that was death. This might be from raking machine-gun fire into the tops of the trees or air-burst HE fragmentation rounds bursting above the tree line from mortars or artillery, but counter-sniping methods were used equally effectively to defeat such threats. The use of flame-throwers at close quarters was also employed to flush out and eliminate the Japanese from their hides in caves, tunnel systems and tree lines where a few well-concealed snipers could inflict a high casualty rate on an advancing force and slow down or even halt an advance across an island until the threat was destroyed.

My point is that it is critical to remain open-minded and focused on the task at hand; that anything and everything must be considered, especially when trying to locate and eliminate your opponent. There is always a first time for something new, slightly adapted or improved and we must be ready in mind and body for this, never dismissing any possibility as you may only ever get one chance. Your opponent will use all his knowledge, skills and ingenuity to outwit and defeat you, including his choice of location.

* * *

I stayed in that position for about two and a half hours with the sentry observing and there was nothing that caught our attention. In my own mind I was convinced that the insurgent had been using the irrigation ditch for movement and to fire from somewhere along this long feature: it was perfect for the purpose. The same ditch also ran flanking the MSG Sanger and in places it was less than 350 metres away from the Sanger location and we knew that he liked to get close. We were now running out of time and light. As planned just at last light Matt and I left our locations, met back up at our small hut and talked about what we had seen, exchanging opinions and discussing our next plan of action.

The next few days seemed to pass by very quickly with a mixture of foot and mobile patrols and a number of contacts against the insurgents as they tried to take on the patrols and our location. In-between patrolling and other taskings all my efforts were spent in hours and hours of observation from different locations around the base over the same area of ground where I perceived this threat to come from. I allowed myself time to observe from different fields of view and at various heights, paying particular attention to the compounds, tracks, ditches and the areas of fresh young crops that were now shooting up and covering the once wet and muddy farmers' fields with almost waist-high plush green vegetation.

Almost any area that once comprised wet mud or just plain soil was now sprouting fast-growing green grass and other forms of vegetation. My arcs of observation were gradually closing down which was bad news for me and good news for the insurgent sniper, giving him more choice in selecting a position to observe and fire from. He had much more cover from view than usual, being screened by all the natural vegetation that was sprouting up everywhere, but we had to locate him as soon as possible.

Insurgent sniper activity was at this time very minimal in our area but active elsewhere with some potentially fatal single shots being fired at friendly forces and luckily for the individuals concerned these were near misses or had impacted somewhere on a helmet or body armour. I thought the sniper had been successful here before and would return to try his luck and prove himself again but Matt and I were still here and waiting for him and our knowledge and understanding of the environmental factors and the ground around us were now even better than before. All that was needed was for us to keep on doing what we had been doing for the past few weeks: waiting and observing, scanning the ground for the slightest indication that he was back and operating in our area; then counter that and act accordingly by locating and destroying the threat as soon as possible.

I had always carried in my kit a small section of folded-up lightweight green wire mesh, just like the mesh you can find in garden centres, and had cut a piece of this large enough to make a false screen for myself. I could then unfold and open it up if I needed to while adopting a prone or variation of this firing position behind in cover to fire from, giving myself the option to fire either with or without the bipod legs attached to my rifle. I had also tied some small thin strips of shredded tan and brown dyed hessian material to it and attached some small thick elastic bands to the green wire so I could fix natural foliage to it as well, adding to the overall camouflage effect. I had also made a crude shroud to cover the front of my objective lens but when the natural light and weather conditions were poor, so was the visibility through it, especially at the further ranges. At least with the wire effect I could widen the hole to allow more light in if needed. I could also unfold the mesh, open it up fully and use it to the same effect as a shroud but with this I was able to mould it over the top of my optical scope and bring it forward to cover the front of my rifle sight, the objective lens coming down at an angle off the top of the rifle scope, breaking up the outline of that area.

I could also open it up to its largest extent and use it to create a false backdrop screen directly behind me if needed, behind my head and shoulders, again with natural foliage attached to it, with a natural gap from which I could observe and fire. I would use this method to cover my rifle scope during the hours of observation and during firing to conceal my scope ring during the middle of the day when the sun was at its highest point and producing the strongest light. This would hopefully conceal the end of the optical sight body and objective lens known as the scope ring, minimizing the risk of any glare, shine or glint from reflected sunlight.

* * *

I had also considered the use of a form of a decoy but in this environment I thought that it was unwarranted. Today's modern optics, even a scope that was only x4 in magnification, at the closer ranges that I knew the enemy sniper preferred were still good enough for him to be able to work out very quickly that the supposed target was simply a decoy and he would not be fooled. The subject of decoys is an interesting one, especially when used in an urban environment in defence or attack to lure out enemy snipers. With experience and understanding of a given tactical situation and in fully understanding the current threat and the overall objectives and aims of a mission or operation, I believe there is a time and place to use such skills. But with today's fast tempo of operations and modern methods of fighting, especially where mobility, speed, firepower and protection may mean that the taking of an objective or neutralizing of a given threat can be achieved in no time at all, combining the co-ordination and use of all military assets gets the job done quickly and effectively with minimal casualties.

As a sniper instructor I believe it is vitally important that an introduction to the use of decoys is taught to young snipers after they have completed their basic sniper course in ongoing continuation training, even if it is just a short interest period in the platoon's training programme. Where possible the snipers should put their ideas into practice out on the field training areas as individuals and as pairs; even in sections working against each other. Then hopefully some lessons will be learned by all those taking part in the field craft lessons and exercises and the good and bad points will be put down to experience for the future.

Decoy tactics are only limited by the resources available to construct such things, your imagination and having the time to construct them, then trying to catch out a fellow trained sniper who is operating in your area. There are some classic old tricks of the trade dating back over a century, such as the making of dummy heads and false uniformed torsos, or the lure of a single torch, candle or lighted cigarette in a position at last light. There is also the use of false firing positions, false crew-served weapon locations or observation posts, command centres or any fake specialist communications equipment for the enemy to find that might lure them in with priority and try to engage such prestige targets, while you and your sniper screen lie in wait to then locate and destroy them. These methods may have worked in conflicts of the past against a similar tactically-trained enemy. However, the threat posed by insurgency – the preferred word used by today's media for our enemy – and the type of fighting conducted by the rebels requires new and old skills and drills to be combined or adapted to meet this current and constantly evolving challenge.

* * *

The natural light during the day was starting to improve and low cloud cover was gradually being replaced by clear blue skies as the sun travelled high across the morning sky and into the early afternoon period. I would try to observe into areas of natural darkness or shadow to see if, when the light passed over any man-made or natural features out on the ground, I could detect just a glimmer of something, a telltale sign that things were not quite right or that a location was possibly being used to conceal something. We knew that the sniper liked to get close, so anything that I thought could be used as a possible firing position from under 400 metres was observed and scanned in great detail. Occasionally I observed out to the further ranges between 400 and 600 metres, or further for possible approach routes into the area, and to prominent points out on the ground within the field of view from my position of observation, which was always where possible slightly elevated from within the confines of the patrol base. I would use the walls and roofs of the outbuildings within the confines of the base with my piece of HESCO wrap to cover myself and my rifle on occasions, especially when operating by the outer wall sections of the MSG Sanger.

However, on this occasion I was not in the prone position on top of one of the Sanger locations or lying out somewhere on the outer but inside the MSG Sanger and had been in there for fifteen minutes or so observing through my binoculars. I was in the standing position trying to squeeze myself into the right corner next to one of the thick main wooden posts supporting the sandbagged roof where the overhead desert camouflaged netting came down off the roof at an angle slightly away from the Sanger itself and was pegged into the ground providing some screening to the sentry. The sentry himself was standing over in the far left corner of the Sanger and observing out over the ground to his front, concentrating on the compounds and the single track that ran very close along the side of the patrol base, past the Sanger and veered over towards the right-hand side of the village. On the other side of this well-used and prominent track were a few compounds that we knew all too well but this area of ground mainly consisted of irrigation ditches, large open flat fields, small areas of dense vegetation and small trees. The wind and light conditions were excellent on this particular day; the best environmental conditions that we had had for a long while, almost perfect.

Natural sunlight, if correctly understood and appreciated, can be an aid to the sniper and used by him to great effect. It can be his friend but also his enemy by possibly catching the sniper out from glare or shine on his optics and also possibly penetrating the camouflage of foliage that he has constructed to create a false screen for cover and used to fire from in his final position. It is also necessary to study and understand the natural tones and shades of colour as light passes over various forms of vegetation and over the ground: this is real attention to detail, and may be necessary against a very experienced and skilled enemy sniper.

The sentry and I were talking and scanning our respective arcs and I was using another pair of binoculars that I had picked up through a friend while on R&R in Germany: they were ex-German army issue and were excellent sturdy optics, Hensoldt Wetzlar 8x30s, replacing my others which had become damaged. I was observing, very slowly scanning over familiar ground with minimal movement, traversing from left to right, overlapping the features on the ground and then stopping at something that caught my eye and studying it in great detail before moving on to the next point of interest, breaking the ground down into even smaller areas for observation.

By now this picture of the surrounding area was imprinted into both Matt's and my head through hours and hours of observation, to the point where I could even close my eyes at night and see the ground and all its detail in colour and picture form in my mind. Our job is not always about fighting with continuous contacts against the insurgents. Any soldier will tell you that there are prolonged periods of time on an operational tour where sometimes whole days, nights or even weeks pass by with nothing happening. Just the normal patrol base routine day in and day out, foot and mobile patrols, and endless hours of sentry duty with plenty of time to observe and assess the ongoing situation or threat against us and come up with possible counter plans.

For me and Matt, it was always a case of watching and waiting for that sole opportunity to locate and PID a threat and destroy that insurgent sniper before he strikes again; once again, that old familiar phrase of enduring patience. The time will come when one of us will foul up, and at the moment he is the hunter and we are the prey so for us it is a matter of just waiting because he *will* come back to us. He is the one who must move around out in the ground where we are observing and find a suitable firing position to use if he wants to take us on, and maybe with a little poor use of ground in his route selection or lack of personal camouflage moving in and out of his position we will be able to locate him. A moment's loss of concentration or idleness and not focusing on the task at hand could result in either of us paying the ultimate price for failure, and hopefully it will not be us.

Ten long days or so had now passed by with the pair of us spending hours and hours observing out over the ground and still nothing. I thought to myself and discussed with Matt that maybe the sniper had moved on out of our area for good. The Sanger locations had not been engaged by single-shot rifle fire for over a week or so, nor had the foot and mobile patrols from our patrol base or the CLPs. After all, he had been successful in other areas around other bases, so maybe was choosing to stay and operate in the new hunting ground where he currently reigned.

Lately when there had been some insurgent sniper activity within our area their initiation was very short and sharp, followed by a heavy rate of small-arms and either RPG or mortar fire onto us. But then all hell usually broke loose, our immediate response being return small-arms fire rapidly followed by the big guns taking over and smashing down some HMG and GMG or even a few artillery rounds from the 105mm light field guns firing from within the base onto identified contact points. Trying to locate the enemy with all this going on took time, and observation was made difficult by the smoke, dust and debris flying all over the place. All this was on the assumption that he was even out there just waiting for the opportunity to strike. The flanks and near distance were my main areas of concentration, although depending on what was going on I would be scanning into the far distances again, watching all vehicle movement and where possible paying attention to the number of people in a vehicle or on the back of a motorbike. Any of these could be the sniper's means of drop-off transport or just transiting around our area to conduct ground and target reconnaissance.

The sun was high and bright above and behind us in our Sanger location and there was hardly a cloud in the sky. The wind conditions were excellent with just a slight breeze that afternoon that we calculated to be an oblique wind which went in our favour, and also that the mirage at the closer ranges was either low or very minimal. The hours were passing slowly that day and as they crept into the early stages of late afternoon the sun had moved across the sky and was now to the side of me as I looked up towards it. Still it was an almost cloudless scene with the light conditions still strong and good.

We had noted a change in the pattern of normal daily life for the locals in and around the compounds over to our right flank, also in the newly-ploughed muddy fields over on the far side of the track and in particular along the main track itself running past our location. These areas were now more or less completely free of the movement of people and animals. Usually at this time of day they would still be working in the fields and transiting up and down the main track on foot or with wooden carts packed to the hilt with crops and pulled along either by donkeys or cattle, very slowly straining under the weight of their cargo. Not even the occasional car or motorbike had passed our location for some time, and the children who would usually be playing outside in front of the compounds were nowhere to be seen. This is usually a good battle indicator: something is about to happen. Over the Sanger intercom came a message from the operations room that another call sign had just come under attack and was taking in a heavy rate of small-arms and mortar fire; also that Fast Air had been tasked and was currently en route.

All of a sudden the unusual eerie afternoon silence was broken. There was frantic activity and noise coming from the area of the gun line: the gunners manning the 105mm light field guns had been immediately stood to, were busy getting to their positions, preparing the ammunition and guns ready to fire and then waiting, poised and ready to receive their first fire mission of the day. I was just scanning the area over to my right and along a familiar irrigation ditch. The foliage here was thick in places and a very dark green in colour as long thin blades of fresh new grass had grown up through the much shorter yellowish dying growth that ran along the top. This ditch followed along the side of another track that could only be seen in a few small places through breaks in the foliage as at ground level it had always been screened by the grass, even through the winter months. Still scanning and ignoring what was going on around me I focused on this particular ditch and the foliage in and around it. It was at this point while using my binoculars that something caught my eye, initially due to shine, glint or glare. On seeing and focusing on this, alarm immediately registered in my brain as it recognized what I was actually observing.

I had seen this many times before while conducting sniper training in which I was the observation post trying to locate snipers and the trainee snipers would have to locate me and another observer in the OP, then to move to a position and try to engage us with blank ammunition without being compromised during the various stages of the test. I myself had caught out many young snipers by observing them due to the front of the objective lens on their rifle scope not being sufficiently concealed to mask the perfect circular dark unnatural ring effect on the lens which may shine or glint in natural sunlight, known as the 'scope ring'. Such lack of attention to detail may mean you can see through the camouflage effect that the sniper has created to conceal himself and then you have them. In training sessions I would describe my observations in detail to the walker [a trained impartial observer] over the radio. The trainee sniper was able to hear what was being said while still fixed in his firing position and it would be explained how and why they had been seen.

The picture that I was observing at the time almost made me want to shit myself: this was fucking real, man-made and a threat to life. It made my heart race as never before, because more or less in every previous engagement I had felt a degree of control. Yet I was the one observing through my optics or rifle scope and in a position to engage at any moment, choosing to a degree where and when to neutralize a given threat and hopefully not the other way round. It looked as though the sniper already had eyes onto this location and was in his final firing position. His efforts at concealment were fucking good but as always we didn't have time to find fault: what we had seen was enough to act.

We tried to remain calm and not draw attention to ourselves by any quick movements in the Sanger. Matt put his binoculars down and switched to the more powerful spotting scope for confirmation as I switched from binoculars to rifle. Everything had been more or less pre-set for a close engagement.

The range had been set on the elevation drum for a possible target at this range, and even if I was slightly out it would only be by a few millimetres either way. As for the wind correction – the setting of the deflection drum – the same principle applied. I was in the standing bipod position using the Sanger parapet for support, leaning as far forward as I physically could and keeping as low as possible for comfortable firing. I adjusted the magnification on my rifle scope and briefly took it up to x20 to observe, then brought it back down to x12 and focused on the observed image. I perceived that to be of a scope ring, a perfect small dark circle, where the sun had just reflected off the front lens of the optic for a moment. Virtually hidden and set back in the long deep dark grass on the forward edge of the ditch, this image filled my sight picture. From what I could make out from his positioning and his line of sight and fire, to this day I firmly believe that the shooter was aiming onto the other Sanger over to our far left and intended trying to take out the sentry in there.

Several moments passed as Matt and I talked to each other about what we were seeing and confirming; he through the spotting scope and I through my rifle scope. I made some final adjustments to my sight picture using the focus control and parallax adjuster and warned the sentry in the Sanger that I was about to fire. There was no time to warn anybody else such as the operations room or the other sentry locations, as for all we knew the sniper could be just about to engage the sentry over in the far Sanger and take him out. I quickly aligned and steadied my sight picture onto the centre of observed mass, then came just slightly away and down from that to roughly the 4 o'clock position. I removed the safety catch and increased the pressure of my hold and grip on the rifle with my right hand, placing my forefinger just gently on the trigger. My left arm was outstretched forward with my left hand holding the area of the bipod where it was attached to the rifle, gently cupping it with my whole hand and pulling it back just slightly so that my firing position was as firm and steady as I could get it. This could be our only opportunity and I could not waste a single second more: I had to fire. Clearing my mind of other thoughts, I concentrated solely on what I was about to do.

I swear my heart had never beat so hard. I was squinting, straining my right eye as I observed through the rifle scope with sweat dripping down the side of my face and onto my cheek piece. I told Matt and the sentry in a very low tone that within three I was going to fire: standby ... standby ... 3, 2, 1, fire! While saying 'Fire!' I simultaneously squeezed the trigger and held that point of aim as still as possible while observing, with just the smallest minute vibration of the sight picture through my rifle scope from my pounding heartbeat and pulse being felt in my hands and feet.

Then came that familiar, sharp distinctive sound ringing out and echoing from the Sanger position as my rifle freed the round from its casing and sent it on its very short journey towards the intended target. I held my position, my whole body frozen, my right eye straining even more through the rifle scope, followed by the smack of my round penetrating something. All I could see was what I perceived to be the scope ring rise up from its present position almost at ground level and go slightly up to the left before disappearing from view. We both waited and watched with anticipation but no movement followed. A few more seconds passed by ... still nothing. As always, while still observing through my scope I operated the rifle bolt, ejecting the empty case slowly, chambered another round ready to fire and held that position. There was a very uneasy silence outside the perimeter of our PB but we could just make out the sounds of a large contact going on over in the far distance. Only a few dark birds had been startled enough to move, taking to the sky from a freshly-ploughed area over in the middle distance in one of the fields to my front, flying away in the direction of the village and heading towards the horizon.

The 105mm light field guns were still silent and their crews still in position ready and waiting to receive a fire mission. Over in the far distance we could now just make out the familiar sound of the Apache helicopters as they arrived on station and started to rain down their deadly selection of firepower from the skies above in support of the call sign in trouble on the ground and hopefully bringing that contact to a swift end and stopping the insurgent attack.

After a few minutes I broke from my position but remained behind my rifle. We were still observing that pinpoint location on the ground and the perceived approaches to it, possible approach routes coming from the left, right and rear that would lead to that position including the general area out to a good few hundred metres. Half an hour passed by with me constantly checking my watch, then an hour, all agonizingly slowly and still with no movement. The only thought in my head at the time was that I needed to confirm this: did I get him? I thought of launching myself over the main wall, across the small ditch and road and legging it, hard-targeting across the open ground as fast as my legs could carry me towards that position with my A2 – and cover from the Sanger which was less than 350 metres away – and confirm the kill. But that would be irresponsible, stupid and reckless. This is not a game, and I quickly put that initial crazy thought away.

The day was now coming to a close and the sun and its natural light were starting to diminish as late afternoon approached leading towards last light. We packed our kit away and went to the operations room to give a full and detailed back-brief on recent events to the boss and Moggy. It was some time later in the evening, about three hours after the event, that the interpreter who had been listening on the ICOM scanner informed us that we had got him, that the insurgent sniper had been missing for several hours, had not reported in and was later found by his comrades under the cover of darkness, dead in his position.

All I can say is that hearing this information and having it confirmed by that means was good enough for me. I simply felt an overwhelming sense of relief that a lot of patience and a little luck had helped us to nail him. Mostly luck, I think, as if the sun had not been so high or bright that afternoon or even in that exact position in the sky, I might not have caught that momentary glimpse of his optic that had drawn us on to him. It was purely the reflection of light from the sun pinging off the optical scope on his rifle while in his final firing position; something I had witnessed before in Afghanistan very early on in the tour. The basic core skills and principles of sniping, the employment of those skills and the ability to adapt and think outside the box are what can keep you alive, together with previous sniping experience and general military operational experience combined, i.e. lessons learned from tour to tour. But whatever it was that day – luck, skill or patience – well it was Endex for him.

But we all knew that he was probably just one of several snipers in that area, that he would be replaced by another and soon forgotten about. And so it would begin all over again: the hunter and the hunted. The cold hard truth, even now in our modern times with our western values, is that being a soldier is just as it has always been throughout the history of conflict since the beginning of time: we are all replaceable and expendable. The struggle between men will continue with even more powerful weapons and greater means to destroy each other. But there will also always be men and women in this world on all sides with a very unique set of natural instincts, skills and certain personal qualities – some acquired through years of dedicated training and experience – and a constant fine-tuning of those skills in the pursuit of excellence. The alternative – failure – is instant and final.

Chapter Twelve

End of Operations

Last Ride to Camp Bastion

We had been waiting for three days for our final lift out by air, our one-way trip by helicopter out of the PB to recover back to Camp Bastion as our tour was coming to its end. The majority of our platoon was now already back there, starting to prepare the vehicles and equipment to be handed over to the next unit who would be taking over the platoon's role in theatre. However, we were stuck out in the patrol base waiting for the weather conditions to improve so we could be extracted out by air. The winds had become very strong and persistent over the last few days and this continued into the late evenings, restricting movement by air in our area.

During the day these were particularly strong and unpredictable as they travelled across the open terrain, whipping up and carrying along the loose fine sand and grit that covered the top layer of the hard-baked soil and gradually picking up momentum. As before, we watched this with interest as what looked like mini tornados would suddenly appear, rising up from a very fine point on the ground with sand and debris being shaped into a swirling plume by the wind, constantly moving and growing wider and wider towards the top while spiralling upwards. First they would travel in one direction, grabbing up any loose foliage and debris that was in their path of travel, and then switch direction in an instant and take off in a totally random manner. Eventually the whirlwinds would disappear out of sight in the far distance or simply die out, dropping their accumulated debris back onto the ground.

The other added hindrance to our departure was the fact that the overall visibility at times was very poor. Even when standing in the centre of the patrol base, when the winds were up you could not observe the inner walls and some of the small outbuildings and Sanger locations that were no further than 100 metres away in places. These were far from ideal conditions to be able to fly or risk an air crew and their airframe for just a routine pick-up.

Over the last week or so the intelligence we had received and our briefings related to the ever-changing situation and the current threat from insurgents within our area which had significantly increased, especially the threat to air. The information geared towards the insurgents wanting to take down an airframe and that they had the resources and capabilities to do this within our area of operations. This would either be by using a vehicle-mounted HMG, meaning the insurgents were more mobile with their firepower, or a variant of some sort of shoulder-launched projectile system to conduct an attack on a helicopter and this was a real threat at the time.

On the third day of sandstorms they started to ease by the middle of the morning and had more or less completely cleared up by late afternoon to reveal an almost clear bright blue sky with the sun in all its glory beaming down bright light and warmth once again and the air and ground temperatures started to rise. By late evening and the darkness of the night the whole sky was crystal clear with what looked like thousands of small white dots littering the sky above us: small white stars just hanging there as if they were watching us. The moon was full and producing enough light so that outside the buildings in the open areas of the patrol base we didn't need to use our head torches to move around at night.

The time had finally come for us to leave and our ops room had received a message just minutes earlier over the radio net that a helicopter was now en route, inbound from Camp Bastion to our location and would be arriving shortly. We said goodbye to our new comrades from 3 Platoon 2 PWRR and some of the Royal Artillery lads and wished them all well. All our bergans and day sacks were stacked up together in single file by the side of the HLS ready to go onto the helicopter when it arrived.

After one last quick visual check of our rifles and equipment and that we now had all four of our H83 containers filled with 5.56mm, 7.62mm or 8.59mm ball ammunition and spare smoke grenades that we had been dragging around with us over the last six months, we were well and truly ready to be extracted off the ground for the final time. Ben, Matt and I waited in silence on the corner of the HLS. We stood by our kit with our rifles slung over our shoulders and our helmets resting on top of Ben's well-packed and oversized day sack, ready to put on as soon as we heard or caught sight of the inbound helicopter. We didn't say much to each other but inside I knew that like me they were excited and relieved that it would soon all be over. Each one of us was somewhere deep in his own thoughts, reminiscing over the last few hectic months, the good and the bad times we had been through and had shared and spent with the lads from 3 Platoon, A Company 2 PWRR, those from 42 Commando J Company and not forgetting our very own platoon, Reconnaissance Platoon 1 PWRR.

This moment of solemn silence and reflection was broken on hearing and seeing the ops room runner dashing towards us. When he reached us he was smiling away and told us that the helicopter was inbound and would be here in about figures five. We put on our helmets and took a knee next to our kit in file, ready to board the chopper: Ben, Matt and me. Then the helicopter was on its final approach, coming in to land on the patrol base HLS, lining up onto the centre and hovering about 60 feet off the ground with the pilot making minor adjustments as it started to descend. The engine noise filled our ears together with the familiar sound of the rotors as they sliced through the air, producing a very powerful downwash of warm air onto us and whisking up all the loose sand and grit on the floor around us into a dust cloud while the strong smell of aviation fuel filled our nostrils.

The helicopter landed and at the same time the loadmaster slid the side door fully open. We waited for him to give us the thumbs-up so we could board but first he had to unload some mail bags for the location which were thrown clear and towards us. We went to grab them and remove them from the HLS, piling them up on one side with the runner. The loadmaster then signalled us, giving us the thumbs-up: this was it, we were finally going. A final quick rush of adrenaline fuelled our bodies as we grabbed our kit and equipment and climbed aboard one by one, helping each other with the kit and throwing it up to the loadmaster who then stacked it behind him. At last we took our seats, putting on the seat belts in a mad rush of excitement.

Once we were all in and secure in our seats the loadmaster slid the door closed and we started to take off, lifting up from the ground and into the darkness. We sat side by side with our rifles held firmly between our legs in the darkness, feeling the constant vibration from the helicopter engine and rotors through our seats and on the floor through our boots as we sat there in silence. The loadmaster had switched to using his night-vision goggles and was constantly looking out of the small windows on the side door to either side of the aircraft. He would spend a short while on one side and then cross over to the other, peering through the Perspex window and every now and then talking to the pilot and navigator through the microphone on his helmet headset.

The journey to Camp Bastion did not take long. We were only in the air for about fifteen to twenty minutes and soon enough could see the bright lights of the massive ever-growing complex of Bastion through the small windows as we approached. We came in low and overflew some buildings of varying size and some large aircraft hangars where a few airframes were parked up on the tarmac runway, looking like they were fuelled, armed and just ready to go. Men and construction vehicles of all sorts could be seen as we overflew them: vehicle lights and men in high-visibility vests scattered in groups working on vast sections of the airfield and newly-constructed buildings, all the time under constant bright light.

As we finally came in to land there was a moment that I shall always remember with crystal clarity. All three of us had undone our seat belts on the final approach to landing, we all looked at each other from under our helmets and all I could see from Ben and Matt were two massive smiles, and with that an excitable conversation started up between us about who was going to do the first NAAFI run. That started up just a few minutes from landing, breaking our silence, and we suddenly stood up while trying to keep ourselves steady on our feet as the helicopter was trying to land. The wheels hit the ground and the chopper steadily came to a halt with the loadmaster sliding the side door fully open. We jumped off one by one in rapid succession, grabbing whatever bergan and day sacks the loadmaster gave us and moved off to the side by about 20 metres until all three of us and our kit and equipment were safely away from the helicopter. As we all took a knee while trying to secure our kit to the ground, I signalled back to the loadmaster that all was okay. He signed back, closing the door, and the helicopter started to rise into the darkness. Once again we felt the force of the downwash, the engine heat and the strong smell of aviation fuel as it powered up to take off.

We all waited for a minute or two for the helicopter to clear the HLS while hunched over on our knees leaning forward onto our day sacks and bergans, waiting with our heads bowed down and holding on to our rifles until the dust and wind settled and then slowly got back up onto our feet. This was it: stand down. For us it was now well and truly over, and all three of us had made it through more or less unscathed. We took off our helmets and undid our body armour on the sides, allowing the cold evening air to get to the warm, damp sweaty areas underneath and what a relief that was. The smiles broke through once more as we shook each other's hands and congratulated each other on making it through to the end of this tour.

Moments later we were sorting out our day sacks and bergans and getting ready to move off when from across the other side of the HLS an American-style civilian pick-up truck arrived, its bright headlights temporarily blinding us till the driver turned away and came to a halt right beside us in front of all our gear. The platoon had tasked the pick-up to come and collect us and lucky for us that it had: we were some distance from our accommodation and it would have been a right old tab across Camp Bastion just to get back to the platoon's quarters laden with all our kit and equipment.

The driver greeted us as he brought the vehicle to a halt and we in turn introduced ourselves while throwing our kit onto the back of the pick-up and climbing in, sorting out the kit so we were able to use it as a comfortable seat in the corners of the truck. When we had finally arranged ourselves in the back, the driver moved slowly away from the HLS and we made our way along the dusty well-lit roads in Camp Bastion, passing men and machines either just deploying out or coming back in. There is always constant activity everywhere both day and night in Bastion, on the ground and in the air: it never sleeps, continually supporting the coalition's ongoing efforts in Afghanistan.

The next day, following a shower, some decent food and a few hours' sleep, it was time for me to crack on with some personal administration. The first task was to de-bomb all my 5.56mm rifle ammunition and the sniper rifle ammunition for my .338 and account for my smoke and HE grenades so that it was all ready to be handed in for the handover to the next incoming unit. As for ball ammunition for the sniper rifle, I had hundreds of extra 8.59mm rounds that I had been given on my travels by other unit CQMS. I had over the last six months collected three H83 containers full of loose and boxed ball ammunition which I hoarded away in my kit, always finding space somewhere either in my bergan, day sack or grip to store it, and the remainder was divided between Ben and Matt to carry. I always had in the back of my mind that I might need it some day and going back to my experiences at CIMIC House in Al Amarah, Iraq, I would say it is better to have it even if you don't use it rather than needing or wanting it and not having it. It was worth the effort of hauling the extra weight around in my kit wherever I went and always having that little extra reserve of 8.59mm ammunition stored away under my camp bed where we lived in the patrol base.

The next task was to prepare other pieces of operational equipment that were issued and used in theatre to be cleaned or handed over to the next unit and the serial-numbered items of equipment to be checked once again and recorded for the flight manifest paperwork, then packed away in the platoon's freight boxes. These were boxed up and secured once the RMP had checked their contents and that there was no contraband – OP plunder, as it was known – of any description hidden or stored away in them. Once checked, the boxes were sealed by the RMP, then secured away ready to be flown out of theatre and back to Germany within the next few days.

As I sat on the roller-track flooring of the tent, cross-legged and slightly hunched forward with my elbows resting on my knees as I cleaned my rifle bipod legs, I could hear Matt's music playing in the background almost in harmony with the generator located outside behind the tent as it hummed away, powering the air-conditioning for the tent and pumping in a continuous flow of cool air. I was surrounded by rifle-cleaning flannelette and lens tissue cloths as well as my lens brushes, with the rest of my rifle-cleaning kit littered over the floor around me. I sat quietly and deep in thought as I attentively and methodically cleaned the rifle, optics and ancillaries for the last time before they had to be put away in a transit box for the trip back to Germany. My .338 was completely stripped down and even the tan tape that covered the barrel and other parts had been removed, leaving a sticky residue in places with ingrained particles of sand and grit which was a pain to remove.

Once I had cleaned my trusty faithful rifle I placed it and all the ancillaries back into its green metal transit box for the very last time in my operational career. I found this very moving as it was like saying goodbye for the final time to a long-standing and trusted friend, effectively laying them to rest.

My rifle had been with me throughout everything and had never let me down; the optics on the rifle had helped me to observe with such clarity; and the whole had helped me to eliminate threats with such precision. I was saying farewell to an old friend: Rifle No1 on the stock, Serial No 12258, Sight No 340095.

For me this was final closure on the end of my operational career as a soldier and military sniper because I had pretty much blagged my way through the hearing tests to be able to deploy to Afghanistan and even then, I was only supposed to be deployed to Camp Bastion in a supporting role for the platoon. My hearing was not what it used to be and any further exposure to sudden loud noises such as explosions and small arms wouldn't be too good for me. But this time I had had the use of much improved hearing defence. So the fact that I had a chance to go out on operations and get out on the ground with the new, improved .338 sniper rifle and maybe could be of some use motivated me intensely to get myself out on operations for the final time in my career with my old reconnaissance platoon.

The end result was that, despite everything, my hearing more or less remained the same. I wore ear defenders and sometimes two pairs when Fast Air and artillery were dropping bombs and pounding positions or an engagement turned into one hell of a fire-fight against the insurgents. I would say that on occasions firing with the suppressor fitted onto the end of my rifle helped as well but that was just my lame excuse. Anyway, it worked and I would not have changed a single thing.

Epilogue

Stalking Shadows

The Man with the Rifle Knows

There is hardly a day goes by where at some point I cannot escape from some of the repressed memories of what I have seen and experienced during my time in the forces. These are normally stored away; hidden memories that can suddenly and randomly come from nowhere to invade my personal thoughts, even if it is just a flicker or a glimpse of a random image for a fraction of a second that disappears just as quickly as it came. Like many other servicemen and women who have experienced or witnessed conflict and suffering, I do see dead people often in my mind: a still picture or colour image of a distant memory associated with or linked to my past; the mental baggage from certain operational tours.

Such things are all stored away somewhere in a very deep, dark place in the furthermost reaches of my memory, stored just as if in an actual physical file. However, the contents have no names or dates, just images; usually of a male head that has received violent trauma, or bloodied bodies. It might be a man's head covered with dark red blood and with small segmented lumps of torn tissue pulp just hanging down from an open wound. Some such images have exposed bone fragments or teeth protruding through the torn flesh remaining around the mouth and jaw line where the round had impacted the soft tissue and penetrated through to the underlying bone structure on entering or exiting the head.

The range to the target, the environmental conditions, the view of observed centre of mass of a target and what this target is doing – static or mobile – determine the threat, and depending on the skill of the firer, a well-aimed and placed round that penetrates through the base of the skull where it joins the neck will ideally pierce the very top of the spinal column and the medulla oblongata, causing immediate death and neutralization of the threat.

Rounds to the torso or main trunk of the body causing instant death usually just leave the fatality in a heap where they fell, either lying prone on the ground or simply slumped over. The clothing and equipment on the body would be heavily stained with blood at the point of impact. The round, having penetrated the soft tissue of the outer skin, entered the torso and on most occasions passed completely through the body, would have caused catastrophic haemorrhage and damage to the internal organs in the bullet's path as it travelled through on its way to exiting the body. Usually behind the target there is a larger exit wound, creating a blood spatter: a pattern of blood including small fragments of a pulpy mixture of tissue and bone, cloth and even dark blood–matted scalp hair. This would stain the wall, vehicle or whatever backdrop was behind the target at the time.

These images were controlled and organized by my subconscious mind into some sort of instantly retrievable colour images, ready to replay just like a DVD in HD, all set to play and fill my mind with a certain image, only with no sound. As soon as I am aware of the trick that my mind is trying to play on me, I try to shut it down straight away by replacing it with the reality of where I am and what I am doing. Most of the time these memories are triggered by sudden loud sounds, certain smells or emotions taking me back to a place and an event that happened a long time ago, just for a second or two. Every now and then while I am asleep these recorded memories or images creep into my unconscious mind but in a much more detailed, violent and disturbing form. These tend to be of longer duration than when I am awake and in control of my mind. I may find myself abruptly awake and fully alert in the early hours of the morning in a nervous sweat with my heart in overdrive going nineteen to the dozen, and thinking 'What the fuck was that all about? And where did it come from?' because sometimes I simply don't recall ever doing or seeing what my mind has just shown me: such sheer violent horror.

Whenever this happens I find myself talking to myself in my head, telling myself that it was just a bad dream, that it is not real, don't think about it even for another moment. I will be trying to reject those images and realize that everything is okay, thinking to myself: 'You fucking glue bag, get a grip of yourself. Go get your kit on and go out for a slow run or bike around the woods. Get some fresh air into your lungs and take in the natural beauty around you. This is real, this is where you are.' I always remind myself of the fact that I am lucky enough to be able to do these things. I do find that exercise in some form or other helps me, especially when it comes to sleeping. Sometimes being a little chinned each day helps me to shut down my mind and my conscious state, along with the tinnitus that is continuously in my head and always with me. I accept the latter and remain positive about it as I am very fortunate to have only that to cope with.

I consider myself to be one of the very lucky ones and now greatly appreciate the smaller things in life that I previously took for granted. I do often look back and think of my actions, some of them in great detail, and always come up with the same conclusion. I work on fact: on what is before me, using my senses and logic, and would not change a single decision that I made as a soldier while serving on operations. I served my Queen and country and took my oath of allegiance to serve and protect the interests of my country and its people. As a soldier serving in a regular infantry battalion, just like the rest of our armed forces personnel in their respective arms of service, I was bound by a unique set of principles found only in such work environments as the military. There is a strong, binding sense of pride in our regiments and we always remember those men and women who have served before us in past times, particularly those who have made the ultimate sacrifice and fallen in conflict so that we may continue to live in our accustomed free lifestyle.

The people with whom we work side by side, day in and day out, really matter and these comrades-in-arms become like an extended family unit. Some of us who have stayed in for the long haul, either from the days of basic training or the early years of our battalion life, and served together side by side in the same regiment, company or platoon since we were 18 years old have grown up while serving together and grown to know each other and our families. This is what I fought for to the best of my ability, so that my comrades, battle buddies and mates would be able to continue the existence that we had before, no matter what was thrown at us. I can live with what I have done but letting down a comrade when they need you – especially when the shit has really hit the fan – well, for me this is just unacceptable and would mean a lifetime of carrying unimaginable guilt.

This means doing what I have been trained to do for so many years and actually having to use those skills, putting them to the test against a fellow human being in conflict. This is it: your time as a soldier has come and this is not some exercise or training drill. The time has come to put all your skills, experience, knowledge and most of all your whole being to the ultimate test. When it all comes down to that definitive moment in time – taking that shot and ending a fellow man's life – can you do it?

Engaging a wooden or metal target out on a range or field firing area in good, clear weather conditions with no mortal threat to yourself makes it relatively easy to do your job during training as you become operationally experienced. In the back of your mind you know that it is not a real combat situation and you know the game plan or how it will run to some degree and the fact that there is no real threat of injury or death lingering around you. However, having to engage a fellow man who is trying to do exactly what you are doing in a hostile environment – some outside factors going against you and sometimes all being in his favour – requires some more personal qualities to help you remain focused and motivated no matter what in order to achieve your aim of destroying the threat to your comrades and yourself.

To put it simply you must remain extremely mentally and physically focused, concentrating solely on the task at hand, how you are going to achieve this task, what support is available to you and fully understanding how best to use it to maximum effect. The ability to think on your own two feet out on the ground when suddenly put in mortal danger and the ability to adapt to that situation or threat in an instant is essential. You must also trust and rely on your own natural instincts; for example, that gut-wrenching feeling in the bottom of your stomach as fear tries to grip you, followed by adrenaline kicking in and taking over your body, and that cold shiver going down the back of your neck with the nervousness of going into a situation. All your senses are working in overdrive, trying to assimilate all the information that the brain is receiving from all your senses working together at the same time. Your brain must try to figure out in a logical manner the nature and location of the threat against you, while still controlling all your actions and reactions when operating in an alerted state fuelled by adrenaline and fear. Trying to stay one step ahead of the game is critical for survival, or the enemy will destroy you given the slightest opportunity. All his efforts are focused on trying to kill or maim you and your comrades and he will do so without pity or regret; just with sheer contempt for you and everything you stand for.

After my initial experiences at CIMIC House in Al Amarah I had to learn to put things away in my mind very quickly as it was all suddenly so very real with real bullets, rockets, mortars and anything else the insurgents could use against us. The age-old rule of conflict when it comes to man against man is either kill or be killed: essentially the basic will to live. Engaging a fellow man in combat is very different from engaging a wooden or metal target: targets don't bleed or fight back, men do.

Personally I have always worked on a matter-of-fact basis: the facts of what I can see, hear, touch and smell. Every single one of us can form our own opinions and views or have strong beliefs and ideals and these are what drives us and commits us down a certain path in life. But to me my conscience is clear: what is done is done, I don't regret anything and I know for certain in my own mind that I would do it all again if I had to.

There are many servicemen and women from our own armed forces as well as our brothers and sisters from the coalition forces who serve together side by side in these hostile places, going up against threats perhaps on a daily basis, whose lives are so precious yet can be taken away or altered forever in an instant by receiving life-changing injuries. It is a fact that the men and women who have received such injuries continue to motivate themselves, going forward and getting better as the days, weeks and months or in some cases years pass by and never look back with regret during the recovery and rehabilitation phases, only forward, setting themselves personal goals to reach and pass. They don't give up, they continue with the fight but in a different way by getting themselves as mentally and physically fit as they can. Every day can be a battle for them but with constant support from their families and friends and overcoming occasional setbacks these are the people who are truly inspirational in many ways, treating life and each new day as a challenge.

For me time continued to move on and pass by very quickly, especially over my remaining three years of service in the army. On leaving Afghanistan I returned to the battalion which was stationed in Germany and where I was to reform the sniper platoon, taking on the role of platoon commander. In this Ben and Matt stepped up as young JNCOs and were instrumental in helping me to build up and reform the platoon, from the selection and training phases of potential soldiers from within the battalion through to emerging fully qualified and badged snipers able to work as a fully-functional platoon or broken down into pairs or a section and attached to the rifle companies within the battalion.

We also worked and trained closely together in qualifying our men on joint exercises out on the field training grounds at Haltern in West Germany which provided excellent terrain for sniper training, alongside the sniper platoon from 5 Rifles who were fully manned and functional under the watchful eye of their SNCO Jonah. This helped to create the foundations for a fully-functional and operational platoon ready to deploy for the next forthcoming tour of Afghanistan.

I then left the platoon as the operational tour training cycle started and as my time was starting to run down I went to the Light Role Training Wing where I was kept busy instructing in the battalion and over at the Queen's Divisional Courses (QDC) which operated out of Sennybridge Camp in Wales. This involved working within a small team of very dedicated and operationally-experienced instructors from other infantry regiments within the Queen's Division. Our role was to select and prepare soldiers to attend either the Section Commanders Battle Course (SCBC) or Platoon Sergeants Battle Course (PSBC) at the Infantry Battle School in Brecon. On completion of these courses, which are deliberately mentally and physically very demanding and arduous, these soldiers from the regiments within our division were then eligible and qualified for promotion to either a full infantry-qualified corporal or sergeant. In these roles they would usually go on after promotion to become either a section commander in a rifle platoon or a platoon sergeant, the second-in-command of a rifle platoon.

From this experience I had many moments during my last fourteen months of service of being very wet, cold and fatigued. For example, I would be running around conducting platoon attacks, ambushes and patrols out on the training areas of Sennybridge in all weathers together with many hours and miles of battle fitness training, pounding the hills in the Brecon Beacons along with men half my age. I was supposed to be slowing down, getting ready for the smooth transition to Civvy Street with my bergan and belt kit thrown in a cupboard somewhere out of sight and mind, not going into Brecon and buying the latest kit and equipment to make life a little more bearable while conducting these young man's physical events and tasks. But I would not have changed it for the world regarding the individuals I met there and hopefully I passed on some of my experience and knowledge to the next up-and-coming generation of soldiers. I found this very satisfying and worth all the effort from 'the old man' as there was still a very strongly-motivated and eager core of soldiers at various stages of their professional military career who wanted to push themselves forward and above all wanted to soldier on combat operations in our regiments from the Queen's Division. These men are the future of such prestigious regiments and hopefully some of these individuals will continue on and strive towards service with Special Forces in the future.

For me, being a part of all this felt like a fitting and justifiable end to the final chapter of my time serving as a soldier: I had made it through to the end. Yet somewhere in the back of my mind there was still a yearning to possibly deploy for just one more little trip to get just one more operational tour as a sniper under my belt, mainly to prove that I could still do it and that I still retained the sense of purpose imbued by my many years of training. I tried to convince myself that I could do it and that single thought lingered in the back of my mind. But then the harsh reality of it all struck home: the fact of the matter was that my body was taking just a little more time to recover from exertion and with regard to my hearing, the medical board would never let me deploy back to Afghanistan or even in a supporting role to Camp Bastion. So it was time for me to accept the inevitability of becoming a civilian and start to prepare myself for that very important time in my life.

It was during this time that the battalion deployed to Afghanistan on Operation HERRICK 12 as part of 20 Armoured Brigade. Ben and Matt were about to deploy back to Afghanistan and they had both completed and passed SCBC and both been promoted to full corporals. Ben had also attended the Sniper Section Commanders Course where he gained a distinction. The tour Operation HERRICK 12 would be a very different experience for the pair of them compared to our last tour together, as by now the IED threat had greatly increased out on the ground, both to the soldier on foot and to the mobile call signs using vehicles. These devices were usually successful for the insurgents and they had switched to this as a favourite means of killing or maiming coalition forces. The IEDs were usually sited in or around VPs, compounds or basically anywhere on the ground where soldiers patrol or have to operate. They also sometimes combined the use of co-ordinated small-arms and RPG fire onto the general area of the device once it had functioned, effectively creating an ambush in their killing area.

The insurgent IEDs were now better constructed for maximum effect and it seemed that more time and thought was put into the placing and concealment of these devices, making them much harder to locate compared to our last tour on occasions where insurgent ground sign awareness was sometimes relatively poor to the trained eye. Our own counter measures, drills and equipment have improved to counter the continually evolving IED threat and are constantly being worked on to reduce its effects on the soldier out on the ground.

Matt returned from his tour in Afghanistan to Germany, has attended the Light Role Reconnaissance Commander's Course and is continuing to soldier on in service in the battalion. He is looking forward in the future to training recruits at the Infantry Training Centre (ITC) and then returning to his sniping career within the battalion, serving in the sniper platoon.

It was while I was mincing [taking it easy] in Germany in the Light Role Training Wing with only a few months of service to go while training up the BCRs (Battlefield Casualty Replacements) for the battalion on tour that I received the unimaginable news that Ben had been very seriously wounded by an IED. He had been put into a deliberate coma so that the medical team could fly him back to Birmingham Hospital for treatment and we all know that when the professional medical staff take those measures to transport a casualty out of theatre, it is not a good sign to put it mildly. He had been injured by an IED blast while going to the aid of a fallen comrade.

All I can say regarding the devastating news about Ben is that in that instant, as soon as I heard his name mentioned and the nature of his injuries, after taking in the information it honestly just hit me. It felt exactly as I imagine the feeling of someone putting a bayonet into your stomach and twisting the blade slowly, wrapping your intestines around it. That feeling was followed by a surge of familiar uncomfortable emotions sweeping over me; the kind that I personally only associate with news of a death and I suppose it is a form of shock. Fortunately Ben is now on the road to recovery and working hard in the rehab process under the watchful eye of all the amazing doctors, nurses, physiotherapists and support staff at Headley Court and making good progress with his recuperation.

As for 'the old man', I am now a civilian. As for the friendships formed while serving in the army and especially the camaraderie between myself, Ben and Matt gained while serving in Afghanistan, those will never go and never fade. That kind of complete trust and respect for each other can only be gained in moments of extreme peril in combat and comes with knowing that if one of us had to make the ultimate sacrifice for another if necessary and if there was no alternative, there would have been no hesitation from any one of us.

Regarding my friendships with other comrades with whom I had the privilege to serve on operations in Iraq going back as far as 2004, that unique connection is still there and remains timeless. Even as we start to grow old and the world changes around us, we brothers remain united by service to the Crown, our country and her people. Together we still retain an unending bond of friendship and loyalty to one another that only those of us who have served together in mortal combat can attain; a sacred bond that holds us all together until our final day on earth.